NEUROMETHODS

Series Editor
Wolfgang Walz
University of Saskatchewan
Saskatoon, SK, Canada

For other titles published in this series, go to
www.springer.com/series/7657

Animal Models of Pain

Edited by

Chao Ma

Department of Anesthesiology, Yale University School of Medicine, New Haven, CT, USA

Jun-Ming Zhang

Department of Anesthesiology, University of Cincinnati College of Medicine, Cincinnati, OH, USA

Humana Press

Editors
Chao Ma, MD
Department of Anesthesiology
Yale University School of Medicine
Cedar Street 333
06520-8051 New Haven
Connecticut
TMP 3
USA
chao.ma@yale.edu

Jun-Ming Zhang, MD
Department of Anesthesiology
University of Cincinnati
College of Medicine
Albert Sabin Way 231
45267-0531 Cincinnati Ohio
USA
Jun-Ming.Zhang@uc.edu

ISSN 0893-2336 e-ISSN 1940-6045
ISBN 978-1-4939-5761-3 ISBN 978-1-60761-880-5 (eBook)
DOI 10.1007/978-1-60761-880-5
Springer New York Dordrecht Heidelberg London

Printed on acid-free paper

Humana Press is part of Springer Science+Business Media (www.springer.com)

Preface to the Series

Under the guidance of its founders Alan Boulton and Glen Baker, the Neuromethods series by Humana Press has been very successful since the first volume appeared in 1985. In about 17 years, 37 volumes have been published. In 2006, Springer Science + Business Media made a renewed commitment to this series. The new program will focus on methods that are either unique to the nervous system and excitable cells or which need special consideration to be applied to the neurosciences. The program will strike a balance between recent and exciting developments like those concerning new animal models of disease, imaging, in vivo methods, and more established techniques. These include immunocytochemistry and electrophysiological technologies. New trainees in neurosciences still need a sound footing in these older methods in order to apply a critical approach to their results. The careful application of methods is probably the most important step in the process of scientific inquiry. In the past, new methodologies led the way in developing new disciplines in the biological and medical sciences. For example, Physiology emerged out of Anatomy in the nineteenth century by harnessing new methods based on the newly discovered phenomenon of electricity. Nowadays, the relationships between disciplines and methods are more complex. Methods are now widely shared between disciplines and research areas. New developments in electronic publishing also make it possible for scientists to download chapters or protocols selectively within a very short time of encountering them. This new approach has been taken into account in the design of individual volumes and chapters in this series.

Wolfgang Walz

Preface

Pain relief has been a major goal of humanity for centuries. A recent survey conducted by National Institute of Health (NIH) estimated that chronic pain syndromes afflicted one-third of the American population and that over 50 million were either partially or totally disabled. As a result of chronic pain, well over 550 million work days were lost, which, together with health-care costs and payments for compensation, litigation, and malpractice, totals nearly $100 billion annually. The cost of human suffering is even greater than the econo mic impact. It is a distressing fact that millions of patients suffering from persistent pain develop serious physical and affective disorders.

For decades, numerous scientists have tried to carry out extensive research to uncover the mystery of pain and to develop effective therapies for reducing pain. Such efforts have significantly improved our understanding of pain and have led to the discoveries of new drugs for pain treatment in humans and animals as well. Our current understanding of pain and the underlying mechanisms of pain have been revealed mostly by experimentation using animal models due to the severe limitations of using human subjects in pain research. We write this book in an attempt to provide readers with a consolidated review of the animal models available for pain research. In preparing this book, we have tried to capture the diversity of animal models that are used to investigate pain mechanisms, ranging from surgical incision to mechanical compression and from spinal cord injury to cutaneous/local inflammation.

Finally, we would like to express our sincere appreciation to all the authors who contributed to this book.

New Haven, CT | *Chao Ma*
Cincinnati, OH | *Jun-Ming Zhang*

Contents

Contributors

MIROSLAV (MISHA) BACKONJA • *Department of Neurology, University of Wisconsin-Madison, Madison, WI, USA*

ALVIN J. BEITZ • *College of Veterinary Medicine, University of Minnesota, St. Paul, MN, USA*

SONIA K. BHANGOO • *Laboratory of Sensory Biology, National Institute of Dental and Craniofacial Research, National Institutes of Health, Bethesda, MD, USA*

TIMOTHY J. BRENNAN • *Department of Anesthesia, University of Iowa Hospitals and Clinics, Iowa City, IA, USA*

MAXIM DOBRETSOV • *Departments of Anesthesiology, Physiology and Biophysics, and Neurobiology and Developmental Sciences, University of Arkansas for Medical Sciences, Little Rock, AR, USA*

BRYAN HAINS • *Department of Neurology and Center for Neuroscience and Regeneration Research, Yale University School of Medicine, New Haven, CT, USA; Rehabilitation Research Center, VA Connecticut Healthcare System, West Haven, CT, USA*

SINYOUNG KANG • *Departments of Anesthesia and Pharmacology, University of Iowa Hospitals and Clinics, Iowa City, IA, USA*

CHAO MA • *Department of Anesthesiology, Yale University School of Medicine, New Haven, CT, USA*

CHOLAWAT PACHARINSAK • *School of Medicine, Stanford University, Stanford, CA, USA*

LAUREN PETTY • *Department of Anesthesia, Indiana University School of Medicine, Stark Neuroscience Research Institute, Indianapolis, IN, USA*

LINTAO QU • *Department of Anesthesiology, Yale University School of Medicine, New Haven, CT, USA*

KE REN • *Department of Neural and Pain Sciences, Dental School and Program in Neuroscience, University of Maryland, Baltimore, MD, USA*

DMITRY ROMANOVSKY • *Departments of Anesthesiology, Physiology and Biophysics, and Neurobiology and Developmental Sciences, University of Arkansas for Medical Sciences, Little Rock, AR, USA*

XUE-JUN SONG • *Department of Neurobiology, Parker Research Institute, Dallas, TX, USA*

CHRISTINA M. SPOFFORD • *Departments of Anesthesia and Pharmacology, University of Iowa Hospitals and Clinics, Iowa City, IA, USA*

JOSEPH R. STIMERS • *Department of Pharmacology and Toxicology, University of Arkansas for Medical Sciences, Little Rock, AR, USA*

LOUIS P. VERA-PORTOCARRERO • *Department of Pharmacology, College of Medicine, University of Arizona, Tucson, AZ, USA*

PAUL W. WACNIK • *Neuromodulation Research, Medtronic Inc, Minneapolis, MN, USA*
KARIN N. WESTLUND • *Department of Physiology, University of Kentucky Medical Center, Lexington, KY, USA*
FLETCHER A. WHITE • *Department of Anesthesia, Stark Neuroscience Research Institute, Indiana University School of Medicine, Indianapolis, IN, USA*
CHAORAN WU • *Departments of Anesthesia and Pharmacology, University of Iowa Hospitals and Clinics, Iowa City, IA, USA*
WENRUI XIE • *Department of Anesthesiology, Pain Research Center, University of Cincinnati Medical Center, Cincinnati, OH, USA*
JUN XU • *Departments of Anesthesia and Pharmacology, University of Iowa Hospitals and Clinics, Iowa City, IA, USA*
JUN-MING ZHANG • *Department of Anesthesiology, University of Cincinnati College of Medicine, Cincinnati, OH, USA*
RUI-XIN ZHANG • *Center For Integrative Medicine, School of Medicine, University of Maryland, Baltimore, MD, USA*

Chapter 1

Assessment of Pain in Animals

Wenrui Xie

Abstract

The assessment of pain is of critical importance for mechanistic studies as well as for the validation of drug targets. The study of pain in awake animals raises ethical, philosophical, and technical problems. Philosophically, there is the problem that pain cannot be monitored directly in animals but can only be estimated by examining their responses to nociceptive stimuli; however, such responses do not necessarily mean that there is a concomitant sensation. In this chapter, I highlight several types of nociceptive stimuli (thermal, mechanical, or chemical), which have been used in different pain models such as acute pain, chronic pain, arthritis pain, inflammatory, and visceral pain. The monitored reactions are almost always motor responses ranging from spinal reflexes to complex behaviors. Most have the weakness that they may be associated with, or modulated by, other physiological functions. The main methods are reviewed in terms of their sensitivity, specificity, and predictiveness. Although the neural basis of the most commonly used tests is poorly understood, their use will be more profitable if pain is considered within the framework of, rather than apart from, the body's homeostatic mechanisms.

Abbreviations

CCD	Charge-coupled device
CNS	Central nerve system
IASP	International Association for the Study of Pain
LEDs	Light-emitting diodes
PWL	Paw withdrawal latency
PWT	Paw withdrawal threshold
NK1	Neurokinin 1
VAD	Vocalization after discharge
VAS	Visual analog scale
VDS	Vocalization during stimulation

Chao Ma and Jun-Ming Zhang (eds.), *Animal Models of Pain*, Neuromethods, vol. 49,
DOI 10.1007/978-1-60761-880-5_1, © Springer Science+Business Media, LLC 2011

1. Introduction

In 1979, the International Association for the Study of Pain (IASP) defined pain as "an unpleasant sensory and emotional experience associated with actual or potential tissue damage or described in terms of such damage" (1). This definition clearly indicates that pain is a multidimensional experience. Pain classified on the basis of its presumed underlying pathophysiology is broadly categorized as nociceptive or neuropathic pain. Nociceptive pain is caused by the ongoing activation of A_δ and C-nociceptors in response to a noxious stimulus (e.g., injury, disease, inflammation). In contrast to neuropathic pain, the nervous system associated with nociceptive pain is functioning properly. Generally, there is a close correspondence between pain perception and stimulus intensity, and the pain is indicative of real or potential tissue damage. Pain arising from visceral organs is called visceral pain, whereas that arising from tissues such as skin, muscle, joint capsules, and bone is called somatic pain. Somatic pain may be further categorized as superficial (cutaneous) or deep somatic pain (Table 1).

Inflammatory pain occurs when the body responds to tissue damage with the release of chemicals from blood vessels, fibroblasts and local macrophages. These pain messengers spread to chemical receptors around the injured area, covering a larger area than the injury itself (hyperalgesia). Additionally, a secondary response occurs when touch-sensitive nerves become involved and light touch begins to cause pain (allodynia). Finally, the central nervous system may become involved with a process called sympathetic coupling and pain may be caused even without light touch (sympathetically maintained pain).

Neuropathic pain is caused by aberrant signal processing in the peripheral or central nervous system (2). In other words, neuropathic pain reflects nervous system injury or impairment. Common causes of neuropathic pain include trauma, inflammation, metabolic diseases (e.g., diabetes), infections (e.g., herpes zoster), tumors, toxins, and primary neurological diseases (3). Neuropathic pain is sometimes called "pathologic" pain because it serves no purpose (3). A chronic pain state may occur when pathophysiologic changes become independent of the inciting event. Sensitization plays an important role in this process. Nerve injury triggers changes in the CNS that can persist indefinitely. Thus, central sensitization explains why neuropathic pain is often disproportionate to the stimulus (e.g., hyperalgesia, allodynia) or occurs when no identifiable stimulus exists (e.g., persistent pain, pain spread). Neuropathic pain may be continuous or episodic and is perceived in many ways (e.g., burning, tingling, prickling, shooting, electric shock-like, jabbing, squeezing, deep aching, spasm, or cold) (Table 2) (4).

Table 1
Examples and characteristics of nociceptive pain (Sources: refs. (48–52))

	Superficial somatic pain	Deep somatic pain	Visceral pain
Nociceptor location	Skin, subcutaneous tissue, and mucous membranes	Muscles, tendons, joints, visceral fasciae, and bones	Visceral organs[a]
Potential stimuli	External mechanical, chemical, or thermal events Dermatologic disorders	Overuse strain, mechanical injury, cramping, ischemia, inflammation	Organ distension, muscle spasm, traction, ischemia, inflammation
Localization	Well localized	Localized or diffuse and radiating	Well or poorly localized
Quality	Sharp, pricking, or burning sensation	Usually dull or aching, cramping	Deep aching or sharp stabbing pain, which is often referred to as cutaneous sites
Associated symptoms and signs	Cutaneous tenderness, hyperalgesia hyperesthesia, allodynia	Tenderness, reflex muscle spasm, and sympathetic hyperactivity[b]	Malaise, nausea, vomiting, sweating, tenderness, reflex muscle spasm
Clinical examples	Sunburn, chemical or thermal burns, cuts, and contusions of the skin	Arthritis pain, tendonitis, myofascial pain	Colic, appendicitis, pancreatitis, peptic ulcer disease, bladder distension

[a]Visceral organs include the heart, lungs, gastrointestinal tract, pancreas, liver, gallbladder, kidneys, and bladder
[b]Symptoms and signs of sympathetic (autonomic) nervous system hyperactivity include increased heart rate, blood pressure, and respiratory rate; sweating; pallor; dilated pupils; nausea; vomiting; dry mouth; and increased muscle tension

Table 2
Examples and characteristics of neuropathic pain (Sources: refs. (48, 49, 53–56))

	Painful mononeuropathies and polyneuropathies	Deafferentation pain	Sympathetically maintained pain[a]	Central pain
Definition	Pain along the distribution of one or multiple peripheral nerve(s) caused by damage to the affected nerve(s)	Pain that is due to a loss of afferent input	Pain that is maintained by sympathetic nervous system activity	Pain caused by a primary lesion or dysfunction of the CNS
Pain characteristics and associated symptoms	Three main types: Continuous, deep, burning, aching or bruised pain Paroxysmal lancinating (shock-like) pain Abnormal skin sensitivity	Quality: burning, cramping, crushing, aching, stabbing, or shooting Hyperalgesia Hyperpathia Dysesthesia Other abnormal sensations	Quality: burning, throbbing, pressing or shooting Allodynia Hyperalgesia Associated ANS dysregulation and trophic changes[b]	Quality: burning numbing, tingling shooting Spontaneous and steady or evoked ± Sensory loss Allodynia Hyperalgesia
Sources	Metabolic disorders (e.g., diabetes) Toxins (e.g., alcohol chemotherapy agents) Infection (e.g., HIV, herpes zoster) Trauma Compressive (nerve entrapment) Autoimmune and hereditary diseases	Damage to a peripheral nerve, ganglion, or plexus CNS disease or injury (occasional)	Peripheral nerve damage (e.g., CRPS II) Sympathetic efferent Stimulation of nerves by circulating catecholamines	Ischemia (e.g., stroke) Tumors Trauma (e.g., spinal cord injury) Syrinx Demyelination
Clinical examples	Diabetic neuropathy Alcoholic neuropathy Postherpetic neuralgia Carpal tunnel syndrome	Phantom limb pain Postmastectomy pain	CRPS Phantom limb pain Postherpetic neuralgia Some metabolic neuropathies	Poststroke pain Some cancer pain Pain associated with multiple sclerosis

ANS autonomic nervous system, *CNS* central nervous system, *CRPS* complex regional pain syndrome types I and II, *CRPS II* complex regional pain syndrome type II, *HIV* human immunodeficiency virus

[a]Sympathetically maintained pain is a pain mechanism, not a diagnosis. It is associated with several types of pain, but it also may exist as a single entity (55)

[b]Focal autonomic dysregulation can manifest with signs and symptoms such as swelling, pallor, erythema (redness), sweating, and temperature changes. Trophic changes include thinning of the skin, abnormal hair or nail growth, and bone changes

To study pain transmission, identify new pain targets, and characterize the potential analgesic profile of novel compounds for pain relief, an array of experimental animal pain models has been developed mainly in rodents, reflecting all types of pain, from acute to chronic, somatic to visceral, and nociceptive to neuropathic, arthritis and cancer-related pain. Depending on the model, pain measurements can encompass spontaneous pain behaviors as well as pain evoked by various modalities (5, 6).

Recent advances in neuroimaging technology have reinforced the concept that the recognition of pain in humans is a multifaceted process that involves the parallel integration of sensory, emotional, and noxious perceptual information by multiple brain structures (7). Similar brain structures are involved in the process of nociception and associated expression of nociceptive behaviors in injured animals (8). Thus, spontaneous and/or evoked nociceptive behaviors in animals are described frequently as either "pain" or "pain-like" behaviors. The absence of verbal communication in animals is undoubtedly an obstacle to the evaluation of pain. The question of pain in animals can be approached only with anthropomorphic references, although differences probably do exist in comparison with humans, notably with respect to certain cerebral structures. Generally, the most reliable signs of pain are physical ones. Zimmermann re-interpreted the IASP definition of pain so that it could be applied to animals: "an aversive sensory experience caused by actual or potential injury that elicits progressive motor and vegetative reactions, results in learned avoidance behavior, and may modify species specific behavior, including social behavior" (9). In contrast with the polymorphic nature of the pain that is described as a sensation in humans, pain in animals can be estimated only by examining their reactions.

Experimental studies on conscious animals are often designated "behavioral studies." The behavioral tests used to study nociception – nociceptive tests – constitute "input–output" systems. As a result, when describing these tests, one must specify the characteristics of the input (the stimulus) and the output (the reaction of the animal).

2. Phasic Pain

These tests are the most commonly used. Usually, these tests rely on an escape behavior/withdrawal reflex or vocalization as an index of pain. The animals have control over the duration of the pain, their behavioral response terminating the stimulus.

2.1. Tests Based on the Use of Thermal Stimuli

In tests involving thermal stimuli, it is always the skin that is stimulated. These tests do not involve visceral or musculoskeletal tissues. The source of nociceptive stimulation can be distant from its target (e.g., radiant heat from a lamp) or can be in direct contact with the skin. Radiant heat constitutes a relatively selective stimulus for nociceptors and has an advantage over the other modes of thermal stimulation in that it produces no tactile stimulus.

2.1.1. The Tail-Flick Test

There are two variants of the tail-flick test. One consists of applying radiant heat to a small surface of the tail. The other involves immersing the tail in water at a predetermined temperature. This test has proved particularly sensitive for studying the analgesic properties of pharmacological substances. It can also be used to evaluate basal thermal pain sensitivity or to study putative genetic differences among animals without drug ("naïve").

2.1.1.1. The Tail-Flick Test Using Radiant Heat

Equipment: Tail-flick analgesia meter; Kim towel or sterilized cloth

The tail-flick test with radiant heat is an extremely simplified version of the method used on human subjects by Hardy et al. (10). Indeed, Hardy and his colleagues eventually used the technique in the rat. Basically, the whole body except the tail of an animal is wrapped with Kim Towel or sterilized cloth (blindfolding will keep the animal from moving violently) taking care not to wrap the animal too firmly. A radiant heat is applied on the tail placed at the specified spot on a *tail-flick analgesia meter.* When the animal feels discomfort it reacts by withdrawal of the tail by a brief vigorous movement (tail flick) (11), which automatically stops the stimulation. The measurement of the animal reaction time, referred to as "tail-flick latency" (period from the beginning of the stimulation until detection of the animal's response) is achieved by starting a timer at the same time as the application of the heat source. By using a rheostat, the intensity of current through the filament and therefore of radiant heat emission can be controlled, so that one can empirically predetermine the time until the withdrawal of the tail. A photoelectric cell stops the timer and switches off the lamp at the moment the tail is withdrawn. Latency is measured three times and taken the median of the data. It is advisable not to prolong the exposure to radiant heat beyond 10–20 s, otherwise the skin may be burned. The advantages of this method are its simplicity and the small interanimal variability in reaction time measurements under a given set of controlled conditions. The tail-flick is prone to habituation, viz., a reduction in the response with repetitive stimulation. This habituation increases with a shortening of the interstimulus interval and with the intensity of stimulation (12).

2.1.1.2. The Tail-Flick Test Using Immersion of the Tail

Equipment: Hot or cold water bath

The use of immersion of the tail is a variant of the test described above. The most obvious difference is that the area of stimulation is far greater. The animals are held firmly over the opening of the bath and their tails submerged approximately half-way into hot water. The nociceptive threshold is taken as the latency of provoking an abrupt movement of the tail and sometimes the recoiling of the whole body. Again, the reaction time is monitored. To minimize damage to the tail, a cutoff is utilized and considered to be the maximum latency. One of the advantages of this method is that the water bath can be set at various temperatures and it can be less sensitive to environmental conditions. However, it requires handling of the animals when testing for behavior, making this measure highly dependent on the experimenter's experience at handling animals. Simply holding the animals (without immersing the tails) over the bath for a few seconds before the first measurement produced more precise baseline measurements; this "training" procedure leads to markedly reduced struggling. This method can also be used to test for reactivity to cold, using a 4 or 10°C water bath and recording latency to withdraw as an index of pain.

2.1.2. The Paw Withdrawal Test

Equipment: Plantar analgesia meter

In principle, this test is entirely comparable to the test of D'Amour and Smith in 1941 but offers the advantage that it does not involve the preeminent organ of thermoregulation in rats and mice, i.e., the tail (13). With the aim of studying hyperalgesic phenomena resulting from inflammation, Hargreaves et al. (13) had an inspired idea for supplementing the model of Randall and Selitto (14): radiant heat was applied to a paw that had already been inflamed by a subcutaneous injection of carrageenin. Basically, the animal moves freely on a glass surface. A focused infrared source is moved under the animal when the animal is not moving, and a button press applies the heat to the plantar surfaces of the foot. When the animal feels the heat and moves the paw, a photosensor stops the clock and shows the latency from heat onset to paw withdrawal. In each test session, each animal is tested in three to four sequential trials at approximately 5-min intervals to avoid sensitization of the response. Some advantages of this method vs. the tail-flick assay are that (1) unlike the scaly skin of the tail, the plantar surface on the foot is sensitive sensory skin, typical of other mammalian skin surface; (2) both paws can be tested and it has been proven a useful behavioral assessment in models of unilateral nerve injury, the contralateral paw serving as control for the injured paw; (3) in this assay, animals are confined in plastic chambers but not manually restrained as in the tail-flick assay or in the immersion test, decreasing the stress level of the test subjects. The test can be improved by minimizing variations in the baseline temperature of the skin such as warming the glass to 30°C to prevent paw cooling and insuring that the tested paw is in contact with the glass (15). However, there is a disadvantage in that the position of the leg becomes a factor since the background level of activity in the flexors varies with the position of the animal.

2.1.3. The Hot Plate Test

Equipment: Hot plate analgesia meter

This test consists of introducing a rat or mouse into an open-ended cylindrical space with a floor consisting of a metallic plate that is heated by a thermode or a boiling liquid up to 65°C. Animals are brought to the testing room and allowed to acclimatize for 10 min before the test begins. Pain reflexes in response to a thermal stimulus are measured using a hot plate analgesia meter. The surface of the hot plate is heated to a constant temperature up to 65°C, as measured by a built-in digital thermometer with an accuracy of 0.1°C and verified by a surface thermometer. Animals are placed on the hot plate, which is surrounded by a clear acrylic cage (open top), and the Start/Stop button on the timer is activated. The latency to respond with either a hind paw lick, hind paw flick, or jump (whichever comes first) is measured to the nearest 0.1 s by deactivating the timer when the response is observed. The animal is immediately removed from the hot plate and returned to its home cage. If an animal does not respond within 30 s, the test is terminated and the animal is removed from the hot plate. Animals are tested one at a time and are not habituated to the apparatus prior to testing. Each animal is tested only once.

A plate heated to a constant temperature produces two behavioral components that can be measured in terms of their reaction times, paw licking and jumping. As far as analgesic substances are concerned, only the paw-licking behavior is affected by opioids. On the other hand, the jumping reaction time is increased equally by less powerful analgesics (16). The specificity and sensitivity of the test can be increased by measuring the reaction time of the first evoked behavior regardless of whether it is paw licking or jumping, or by lowering the temperature. The behavior is relatively stereotyped in the mouse but is more complex in the rat, including sniffing, licking its forepaws or hind paws, straightening up, stamping its feet, starting and stopping washing itself. Because so many of these behaviors exist, observation of them is difficult. Furthermore, this test is very susceptible to learning phenomena, which result in a progressive shortening of the jumping reaction time accompanied by the disappearance of the licking behavior. Thus, the animal may lick the paws and then jump during the first test but will jump almost immediately during subsequent tests (17). Similarly, even putting the animals on an unheated plate just once to watch the test leads in subsequent tests to a diminution in the reaction time under standard conditions with a constant noxious temperature (18). Finally, reiteration of the test once a day or once a week inevitably leads to a progressive decrease in the reaction time (19). All these factors make this test a very delicate one to use.

2.1.4. Tests Using Cold Stimuli

Equipment and reagent: cold plate analgesia meter; cold water analgesia meter; acetone

Cold is very rarely used to test acute pain. On the other hand, it is more common to test cold allodynia in animal models of neuropathies. The techniques are directly inspired by those that use heat by contact: immersion of the tail or a limb (20), or placing the animal on a cold surface (21), a cold plate cooled by cold water circulating under it. The temperature (−5 to 25°C) of the cold plate, which is equipped with a Plexiglas box to contain test animals, is set and allowed to stabilize for 5 min (ambient temperature of testing room 21 ± 1°C). The animal is then placed onto the cold plate and the time taken for the first brisk lift or stamp of the ipsilateral hind paw to occur is recorded. Locomotor movements are quite distinct, involving coordinate movement of all four limbs, and these are excluded. The time to the brisk response is interpreted as the latency for cold pain withdrawal. A maximum cutoff time of 150 s is used to prevent tissue damage at the lower temperatures. Each animal is only tested once on any given test day to avoid any possible anesthetic or tissue damage effects that could be produced by repeated exposure to a cold surface. As for the hot plate assay, the cold plate test has the advantage of not necessitating animal restraint. However, depending on the position of the animal paw on the plate (or just above it) the cold stimulation can be very variable. In addition, another limitation of this method is in the case of "whole body" neuropathies such as observed in diabetic neuropathic animals or following chemotherapy. In these models, animals have allodynia in all four paws and the use of a cold plate assay is very difficult if not impossible.

Another more widely used method to test cold sensitivity is putting a drop of acetone on the plantar skin of animals resting on an elevated mesh floor. Basically, animal is placed in cages with a metal mesh floor. After habituation, 50 μL acetone is vaporized on the plantar surface of the paw. The total duration of paw withdrawal, defined as the total time of flinching, licking or biting of the limb, is measured over 30-s to 5-min test period for each of acetone application. Acetone produces a distinct cooling sensation as it evaporates. Normal rats will not respond to this stimulus or will give a very small response (in amplitude and duration), while nerve-injured rats will almost always give an exaggerated response.

2.2. Tests Based on the Use of Mechanical Stimuli

2.2.1. Randall and Selitto Test

Equipment: Analgesy meter for the rat paw (Randall–Selitto); rodent pincher

The preferred sites for applying nociceptive mechanical stimuli are the hind paw and the tail. A common way to assess for acute mechanical sensitivity is using withdrawal threshold to paw/tail pressure using the Randall–Selitto test (14). The *analgesy meter for the rat paw* allows for the application of a steadily increasing pressure to the dorsal surface of the rat's hind paw, tail or muscle via a blunt point (dome-shaped plastic tip) mounted on top of a system of cogwheels with a cursor that can be displaced along the length of a graduated beam. These devices permit the application of increasing measurable pressures and the interruption of the test when the threshold is reached. The measured parameter is the threshold (weight in grams) for the appearance of a given behavior. In practice, the animals are restrained around the trunk with a towel to calm them, and treated gently during the experiments. A cone-shaped pusher with a rounded tip is applied to the paw, tail, or muscle through shaved skin. The rate of force application is set and there is a cutoff loading to avoid damaging the tissue. When the pressure increases, one can see successively the reflex withdrawal of the paw, a more complex movement whereby the animal tries to release its trapped limb, then a sort of struggle, and finally a vocal reaction. The intensity of pressure causing an escape reaction is defined as the withdrawal threshold. The threshold (in g) for either paw/tail withdrawal or vocalization is recorded. It is worth noting that training the animal helps in obtaining a more stable response with this assay. Training sessions are carried out for several consecutive days to increase the sensitivity of the test.

2.2.2. Pricking Pain Test

Equipment: Rodent pincher

Another approach to test for mechanical sensitivity is to use a pinprick, applying painful pressure to the plantar surface of the hind paw. This is similar to the pricking pain test done during the neurological exam in patients and represents an alternative to the "Randall and Selitto" test. In practice, the animal is gently restrained and maintained in a natural position. The force is applied between the two tips of a rodent pincher and is independent of the movements of the limb. The rodent pincher displays the force at which the animal reacts, and reports the mechanical nociception threshold. The behavior can be measured by the duration of paw lifting following the pinprick application or recorded as a frequency of withdrawal (% of response to the pinprick in ten trials).

2.2.3. Von Frey Test

Equipment: Von Frey monofilaments

Finally, mechanical hypersensitivity can also be tested with von Frey monofilaments. The von Frey filament test, developed more than 100 years ago, is still widely used today for the assessment of tactile allodynia. Von Frey monofilaments are short calibrated filaments (nylon filaments are mainly used today) inserted into a holder that allows the investigator to exert a defined pressure on a punctiform area of the rodent paw. The animal is repeatedly stimulated with increasingly stronger filaments to determine the threshold where a nocifensive paw withdrawal response is reliably elicited. A major disadvantage of this test lies in the fact that low-threshold mechanoreceptors are also stimulated, thus the stimulus is not specific. As a consequence, it is difficult to determine in healthy tissues, whether the withdrawal response is triggered by a pain sensation. It is also very difficult to apply the stimulus to a freely moving animal. If tactile allodynia is being measured, animals are placed in a plastic cage with a wire mesh bottom which allows full access to the paws. The area tested is the mid-plantar hind paw, in the sciatic nerve distribution, avoiding the less sensitive tori (footpads). The paw is touched with one of a series of eight von Frey hairs (nylon monofilaments of various thicknesses) with logarithmically incremental stiffness (0.41, 0.70, 1.20, 2.00, 3.63, 5.50, 8.50, and 15.10 g). The von Frey hair is presented perpendicular to the plantar surface with sufficient force to cause slight buckling against the paw, and held for approximately 6–8 s. Stimuli are presented at intervals of several seconds. A positive response is noted if the paw is sharply withdrawn. Flinching immediately upon removal of the hair is also considered a positive response. Based on observations on normal, unoperated rats and healed, sham-operated rats, the cutoff of a 15.10 g hair (~10% of the body weight of the smaller rats) is selected as the upper limit for testing, since stiffer hairs tend to raise the entire limb rather than to buckle, substantially changing the nature of the stimulus. The 50% withdrawal threshold is determined using the up–down method of Dixon (22). In this paradigm, testing is initiated with the 2.0 g hair, in the middle of the series. Stimuli are always presented in a consecutive fashion, whether ascending or descending. In the absence of a paw withdrawal response to the initially selected hair, a stronger stimulus is presented; in the event of paw withdrawal, the next weaker stimulus is chosen. According to Dixon, optimal threshold calculation by this method requires six responses in the immediate vicinity of the 50% threshold. Since the threshold is not known, strings of similar responses may be generated as the threshold is approached from either direction. The resulting pattern of positive and negative responses is tabulated using the convention, X = withdrawal; 0 = no withdrawal, and the 50% response threshold is interpolated using the formula: 50% g threshold = $(10^{(X_f+\kappa\delta)})/10{,}000$, where X_f = value (in log units) of the final von Frey hair used; κ = tabular value for the pattern of positive/negative responses; and δ = mean difference (in log units) between stimuli. Thresholds thus computed do not yield a mathematical continuum (not all possible values can be generated); thus, these results are considered to be nonparametrically distributed (23).

3. Arthritis Pain

Arthritis is the inflammation of a joint, which can include infiltration of inflammatory cells (monocytes), synovial hyperplasia, bone erosion and new bone formation, narrowing of the joint space, and ankylosis of the joint. The most common form of arthritis is osteoarthritis. Pain in osteoarthritis is localized and use-related, occurring during movement or weight bearing. Rheumatoid arthritis is an autoimmune disease of the synovium that leads to an inflammatory poly-arthritis, and is characterized by the symmetrical pattern of affected joints and by morning stiffness, joint swelling and tenderness. Pain in rheumatoid arthritis improves with movement. Gout represents one of the most painful forms of arthritis. The metatarsophalangeal joint (big toe) is typically affected, but other joints can be involved as well, including the knee. Animal models have been developed to investigate the pathophysiology of different forms of knee joint arthritis and are used for the assessment of joint pain.

3.1. Pain Behavior of Arthritic Animals

The main challenge of assessing knee joint pain has been to develop tests that actually measure the sensitivity of the knee joint rather than that of the hind paw. Behavioral tests that use indirect measures of knee joint pain in arthritis models include static and dynamic weight bearing; foot posture and gait analysis (24), including paw elevation time during walking; spontaneous mobility (25); and mechanical or heat sensitivity of the paw (26, 27). Though indirect measures, weight bearing, and gait analysis have the advantage that they are also used in the clinical setting to assess pain in patients with arthritis. More recently, behavioral tests have been developed that directly assess the mechanical sensitivity of the knee by measuring the hind limb withdrawal reflex threshold of knee compression force, struggle threshold angle of knee extension, and vocalizations evoked by stimulation of the knee.

3.2. Weight Bearing

Equipment: Incapacitance tester; CatWalk setup

A significant shift of weight from the arthritic site to the contralateral limb, i.e., a weight-bearing deficit, is taken as a pain measure and has been shown in knee joint arthritis models (24). Measurements of weight bearing are used in arthritis models. Most commonly, the weight distribution on the two hind paws is measured as the force exerted by each limb on a transducer plate in the floor over a given time period (24). Rats are carefully placed in an angled Plexiglas chamber positioned so that each hind paw rests on a separate force plate. Care is taken to ensure that the animal weight is directed onto the force plates and not dissipated through the walls of the chamber. The force exerted by each hind limb (measured in grams) is averaged over a 5-s period. Weight borne by each hind limb is expressed as percent of body weight or percent of weight borne by both hind limbs. The ratio or difference of weight distribution (force) between each hind limb is also calculated. These static measurements of weight bearing by the hind limbs typically involve restraining the animals and do not assess the shift of weight distribution to the forelimbs as occurs with hind limb arthritis. Weight bearing across all four limbs is also measured in rats with knee joint arthritis. Weight load on each limb is detected while the animal is walking across four pairs of force sensor plates in the floor of an enclosed walkway. The digitized output and simultaneously videotaped images are used to calculate the peak vertical load bearing by each limb. Weight distribution across the four limbs is also determined with a gait analysis system ("CatWalk") that measures the intensity of the illumination caused by paw contact with a glass floor (28). The animals traverse a walkway (plexiglass walls) with a glass floor located in a darkened room. Light from a completely encased white fluorescent tube enters the distal (from the observer) long edge of this glass floor. Sufficiently far from the edge, it strikes the surface below the critical angle and is entirely internally reflected. Only at those points where a paw touches the glass, light exits the floor and scatters at the paw, illuminating the points of contact only. Via a mirror, the corridor's floor is monitored by a CCD camera equipped with a wide angle objective. A potential problem with dynamic weight-bearing measurements is that animals are required to move, which can be influenced by a number of factors such as motivation.

3.3. Posture and Gait Analysis

Equipment: Stainless steel cylinder

Related to the assessment of weight bearing, abnormal posture of the hind paw and gait are quantified in knee joint arthritis models using subjective rating scales. Static (standing) and dynamic (walking) behaviors are analyzed separately to calculate a "pain score" in rats. Categories of the rating scale include complete touch of foot pad, partial touch or one foot stand (standing position) and slight limping, severe limping or one foot gait (walking state). A combination of posture and gait analysis is used to rate pain related spontaneous behavior in the knee joint arthritis (29). Behavioral signs include curling toes, eversion of the foot, partial weight bearing, non-weight bearing and guarding, and avoiding contact with the limb. Gait disturbance is detected through measuring increased paw elevation times in arthritic rats walking on a rotating mesh-covered steel drum. An electrode is attached to the plantar surface of each hind paw between the plantar pads. Rats are placed on a stainless steel cylinder of 30 cm in diameter. The cylinder is rotated at 4 rpm forcing the rats to walk. The time of contact between each of the rat's hind paws and the cylinder is measured. When the electrode placed on the animal's paw makes contact with the cylinder floor, a circuit is closed and the time that the circuit remained closed is recorded. Paw elevation time or the ratio of time of contact of the affected foot and the control foot serve as indicators of pain-related functional impairment. The advantage of this gait analysis test is that the quantization (paw elevation time) is independent of the observer.

3.4. Spontaneous Mobility

Equipment: Biotelemetry system; activity boxes

Loss of spontaneous mobility has been detected in rats with knee joint arthritis, presumably related to pain. Locomotor activity is measured in arthritic rats using biotelemetry or activity boxes. The biotelemetry system comprises a transmitter implanted in the peritoneal cavity of the rodent, and a receiver beneath the cage. The signals from the transmitter, which include the body temperature and locomotor activity of the animals, are relayed by a consolidation matrix into a peripheral processor. The receiver detects the radio waves and activity of the rodents as counts, which are registered in the computer system (25). Spontaneous exploratory activity is measured using activity boxes that are divided into zones by photobeams consisting of pairs of infrared light-emitting diodes (LEDs) and phototransistors. Frequency and pattern of photobeam interruption by the animal's movements are recorded on a computer.

3.5. Mechanical or Heat Sensitivity of the Paw

Equipment: Von Frey filaments; analgesy meter for the rat paw (Randall–Selitto); hot plate

Rats with knee joint arthritis have decreased paw withdrawal threshold (mechanical allodynia) and thermal paw withdrawal latency on the affected limb (30). Von Frey filaments and a modified Randall–Selitto analgesiometer are also used to assess the mechanical sensitivity of the hind paw in animals with knee joint arthritis. Thermal sensitivity of the paw in arthritic rats is measured using the hot plate test and the paw withdrawal latency to noxious heat. (Please see previous section for the detailed methods.)

3.6. Mechanical Sensitivity of the Knee

Equipment: Calibrated forceps

The threshold for hind limb withdrawal reflexes evoked by compression of the knee decreases in the arthritic knee in rat and mouse arthritis models (31) (32). Before testing, animals are trained to stay in a restraining device three times a day for 2 days. While the animals are restrained, the experimenter extends one hind limb, and the muscle or knee joint is compressed by calibrated forceps equipped with force transducers (strain gauges), digitized and recorded on a computer and/or displayed in grams. The contact area of the forceps is approximately 30 mm^2. Compression is stopped when the animal withdraws the limb forcefully or when it vocalizes. Hind limb withdrawal reflexes are assessed by scoring the intensity of the manual compression of the knee required to evoke the reflex.

3.7. Struggle Threshold Angle of Knee Extension

In rats with knee joint arthritis the struggle threshold angle of the extension of the arthritic knee is decreased compared to the normal knee. Reduced range of motion and mechanical sensitivity of the arthritic knee is assessed by measuring the struggle threshold of the knee extension angle (33). The animals are gently restrained by one hand to measure the struggle threshold of knee extension. While holding the animal in the palm of the experimenter's hand, the thigh is fixed by holding it with the thumb and the first finger of one hand. Using the fingers of the other hand, the tibia is extended until the animal struggles. To do this, the distance that the heel of the foot travels during the extension is measured. The extension angle is then calculated by trigonometric functions using the length of the tibia and the foot travel distance during extension.

3.8. Vocalizations Evoked by Compression of the Knee

Equipment: Recording chamber

Rodents vocalize in the audible and ultrasonic ranges. When evoked by noxious stimuli, audible vocalizations represent a nocifensive reaction whereas ultrasonic vocalizations in the 22 kHz range reflect an emotional-affective response (34). Vocalization thresholds are significantly decreased in knee joint arthritis models (33). The threshold of audible vocalizations is measured by compressing the knee of manually restrained rats with a calibrated forceps as described above. A recording chamber and computerized analysis system has been developed to measure simultaneously audible and ultrasonic vocalizations evoked by stimulation of the knee (34). Audible and ultrasonic vocalizations are measured simultaneously before and after stimulation. Recordings are made with a condenser microphone (audible range: 20 Hz–16 kHz) connected to a preamplifier and with a bat detector (ultrasonic range; 25 ± 4 kHz). The microphone and bat detector are placed on a platform in front of the animal at a fixed distance (6 cm). The recorded signals are filtered and amplified and fed into a personal computer. Experiments are carried out in a quiet area and appropriate filtering levels are used to avoid the recording of any background noise. Vocalizations are recorded for periods of 2 min. Rate and duration of audible and ultrasonic vocalizations are increased in rats with a knee joint arthritis (34). Vocalizations that occur during stimulation (VDS) and vocalizations that outlast the stimulus (vocalization after discharges, VAD) are analyzed separately. VDS are organized in the brainstem at the medullary level whereas VAD are organized in the limbic forebrain, including the amygdala.

4. Tonic and Visceral Pain

Thermal and mechanical noxious stimuli are usually of short duration and confined to the area of the skin that has been exposed to the stimulus. However, most of the clinically relevant pain states involve pain that arises from deep tissues and visceral pain. These forms of pain are often poorly localized or diffuse and may radiate over considerable distances. The neuronal processes underlying deep and visceral pain are thought to be considerably different from those associated with cutaneous pain. In the following sections, we will describe the assessment of visceral and deep pain with the writhing test and the formalin test.

4.1. Writhing Test

Reagents: glacial acetic acid (0.3–0.6%), 2-phenyl-1,4-benzoquinone (0.02% in 5% aqueous ethanol), and magnesium sulfate.

If a chemically noxious substance is injected into the peritoneal cavity it may activate nociceptors directly and/or produce pain through inflammation of visceral (subdiaphragmatic organs) and subcutaneous (muscle wall) tissues (35). The most commonly used substances for the stimulation of visceral pain in mice are glacial acetic acid (0.3–0.6%), 2-phenyl-1,4-benzoquinone (0.02% in 5% aqueous ethanol), and magnesium sulfate. The test is sometimes called the abdominal contortion test, the abdominal constriction response, or the stretching test, but more commonly it is known as the "writhing test."

For the assay, animals are placed in a small observation chamber (e.g., an animal cage) and habituated for at least 10 min. The noxious substance is injected into the peritoneum in a volume of 10 ml/kg. Within minutes after the injection, a typical "writhing" response, indicative of visceral pain, can be observed. The intraperitoneal administration of agents that irritate serous membranes provokes a very stereotyped behavior in the mouse and the rat which is characterized by abdominal contractions, movements of the body as a whole (particularly of the hind paws), twisting of dorsoabdominal muscles, and a reduction in motor activity and motor in coordination. Generally, the measurements are of the occurrence per unit of time of abdominal cramps resulting from the injection of the algogenic agent for 15–30 min. Immediately after the conclusion of the test, the animal should receive an injection of an analgesia drug such as buprenorphine.

Writhing responses are considered to be reflexes (36) and to be evidence of visceral pain (37) . Unfortunately, the frequency of cramps decreases spontaneously with time to such an extent that it is impossible to evaluate the duration of action of an analgesic on a single animal. Furthermore, the number of cramps is subject to a great deal of variability.

After first described in 1957 by Siegmund et al. (35) who performed intraperitoneal injection of phenylbenzoquinone to observe the analogous effect following intraperitoneal injection of radio-opaque elements, the writhing test has been modified many times. These modifications mainly concern the chemical agent that, in turn, determines the duration of the effect: acetylcholine, dilute hydrochloric, or acetic acid, bradykinin, adrenaline, adenosine triphosphate, potassium chloride, tryptamine, and ocytocin have all been used. Modifications have also been made to the concentration, temperature, and volume of the injected solution, the experimental conditions, and ways of monitoring behavioral changes so as to simplify the test and increase its sensitivity (38).

4.2. Formalin Test

Reagents: formaldehyde; hypertonic saline; ethylene diamine tetra-acetic acid (25 mg/ml in saline); Freund's adjuvant; capsaicin; bee venom

The formalin test was originally described by Dubuison and Dennis (39) using rats and was later modified for use in mice (40). The formalin test is an important animal model in the study of acute long-lasting pain. The term *formalin* usually means a 37% solution of formaldehyde. Less commonly used are hypertonic saline, ethylene diamine tetra-acetic acid, Freund's adjuvant, capsaicin, and bee venom. Depending on the specific goal of the experiment, formalin can be injected into different body regions, either subcutaneously or intramuscularly. In this test, most commonly, a small volume of 5% formalin is injected under the dorsal surface of the plantar surface of the hind paw. For analysis of pain-evoked responses the treatment should be limited to a single injection in the back of the hind paw, for a number of reasons, specifically the forelimbs are often used in grooming behavior; it is easier to inject formalin into the soft cutaneous tissue of the back of the hind paw, as opposed to the ventral paw or the ankle; and the animal's walking is not modified by the presence of the fluid in this site. Apart from the concentration, the specific dose of formalin varies among laboratories and according to the objectives of the experiment. The average dose is 10–20 μl for mice and 50 μl for rats, although in rats doses of 80–150 μl have often been used, and in some cases it has been as high as 250 or 400 μl. The decision to use high doses must be carefully evaluated.

Formalin-induced pain evokes three main behavioral responses: licking, tonic flexion, and phasic flexion of the injected limb ("paw jerk"), the frequency, duration, and level of which depend on the specific concentration used and the site of injection. All pain-evoked responses appear immediately after treatment and disappear within 1–2 h, depending on the concentration, although the swelling induced by the inflammation can last for several days. In the rat and the mouse, intraplantar injections of formalin produce a biphasic behavioral reaction. This behavior consists of an initial phase, occurring about 3 min after the injection that may last for 5–10 min. After this early phase, the animals show relatively few pain responses for 5–10 min (quiescent period), until the second phase response begins between the 20 and 30 min that may last for up to 30 min (41). The intensities of these behaviors are dependent on the concentration of formalin that is administered. The first phase results essentially from the direct stimulation of nociceptors, whereas the second involves a period of sensitization during which inflammatory phenomena occur.

Prior to the test, the animals are habituated to the observation chamber for 10 min. The chamber should have a transparent floor with a mirror mounted underneath to allow a clear view of the paws. Early phase responses are observed in the 1–10 min interval following the injection and late phase responses 20–40 min after the injection. The nociceptive responses to formalin are commonly assessed by the weighted-scores method of behavioral rating (42). Formalin-evoked painful behavior can be assessed on a four-level scale related to posture: 0, normal posture; 1, with the injected paw remaining on the ground but not supporting the animal; 2, with the injected paw clearly raised; and 3, with the injected paw being licked, nibbled, or shaken (39). The response is given a mark, and the results are expressed either continuously per unit of time or at regular time intervals when several animals are observed sequentially. The measured parameter can also be the number of licks or twitches of the paw per unit of time, the cumulative time spent biting/licking the paw, or even a measure of the overall agitation of the animal obtained by a strain gauge coupled to the cage. Such specific behaviors resulting from an injection of formalin can be captured automatically by a camera attached to a computer; in this way, the effects of a pharmacological substance on such motor activity can be identified, analyzed, and uncoupled from antinociceptive effects. Immediately after the conclusion of the test, the animal should receive an injection of an analgesic drug such as buprenophine. It should be noted that the formalin injection will produce a small necrotic area which may require 7–10 days to heal.

5. Spontaneous Pain

For patients with chronic pain, the personal description of pain, based on either verbal report, diagrammatical representation of cutaneous spread, completion of pain questionnaires such as the McGill Pain Questionnaire, and pain scales such as the visual analogue scale and neuropathic pain scale, provide health specialists with information about the intensity, duration, and location of the pain. While we cannot ask an animal directly about the ongoing nature of its pain experience, many of the behaviors that are thought to represent spontaneous pain have been reported in different animal models of persistent, inflammatory and neuropathic pain.

In models of neuropathic pain (43), the repertoire of behavioral changes that indicate spontaneous pain include increased weight bearing on the uninjured hind limb and guarding behavior of the injured paw that endure for several weeks after the induction of injury. Additional behaviors can include licking the injured paw together with "gentle" skin biting and pulling the toenails. Autotomy is also seen in rodents after complete sectioning of the sciatic and saphenous nerves, which is associated with a permanently anesthetised foot (44). However, although such behaviors can be interpreted as pain-like behavior, they might also represent negative signs such as *paresthesiae* and *dysethesiae* (tingling and numbness). These negative neuropathic pain signs of dysethesiae and/or paresthesiae are the most common sensory complaints reported by patients with clinical peripheral neuropathies, rather than pain per se (45).

6. Influence of Species and Genetic Line

We must always bear these factors in mind because they can influence the pharmacokinetics and pharmacodynamics of administered substances just as much as the physiological mechanisms that underlie the recorded responses. In this context, the study of ten lines of mice subjected to a series of different tests of nociception revealed a strong genetic influence on the responses of the animals; for example, one stock of animals showed virtually no responses to the formalin test (46). Similarly, in the context of the hypothalamo-hypophyseal axis, the responses to stress vary according to the stock of rats, which have low or high sensitivities. This results secondarily in the opposite susceptibility for inflammatory diseases (47). Variability can also relate to the anatomy of the nervous system: noradrenergic neurons from the locus coeruleus project toward the dorsal or ventral horn, depending on whether Sprague–Dawley rats belong to the Harlan or the Sasco stock (48). At a pharmacological level, the effects of morphine are also genetically determined, at least in the mouse (49). Interspecies variability is undoubtedly even greater. For example, NK1 receptors in humans are identical to those in the guinea pig but different from those in the rat and mouse (50). The pharmacological effects can also vary radically from one animal species to another. Veterinarians have known for a long time that the properties of morphine vary radically with species.

References

1. Pain terms: a list with definitions and notes on usage. Recommended by the IASP Subcommittee on Taxonomy. Pain 1979;6(3):249.
2. Portenoy RK. Neuropathic pain. In: Portenoy RK, Kanner RM, eds. Pain Management: Theory and Practice. Philadelphia: FD Davis; 1996:83–125.
3. Woolf CJ. Pain. Neurobiol Dis 2000;7(5): 504–10.
4. Portenoy RK, Kanner RM. Definition and assessment of pain. In: Portenoy RK, Kanner RM, eds. Pain Management: Theory and Practice. Philadelphia: FD Davis; 1996:3–18.
5. Colpaert FC. Evidence that adjuvant arthritis in the rat is associated with chronic pain. Pain 1987;28(2):201–22.
6. Kauppila T. Correlation between autotomy-behavior and current theories of neuropathic

pain. Neurosci Biobehav Rev 1998;23(1): 111–29.
7. Rainville P. Brain mechanisms of pain affect and pain modulation. Curr Opin Neurobiol 2002;12(2):195–204.
8. Chang C, Shyu BC. A fMRI study of brain activations during non-noxious and noxious electrical stimulation of the sciatic nerve of rats. Brain Res 2001;897(1–2):71–81.
9. Zimmermann M. Behavioural investigations of pain in animals. In: Duncan IJH, Molony Y, eds. Assessing Pain in Farm Animals. Bruxelles: Office for Official Publications of the European Communities; 1986:16–29.
10. Hardy JD, Wolff HG, Goodell H. Studies on pain. A new method for measuring pain threshold: observations on spatial summation of pain. J Clin Invest 1940;19(4):649–57.
11. d'Amore A, Chiarotti F, Renzi P. High-intensity nociceptive stimuli minimize behavioral effects induced by restraining stress during the tail-flick test. J Pharmacol Toxicol Methods 1992;27(4):197–201.
12. Carstens E, Wilson C. Rat tail flick reflex: magnitude measurement of stimulus-response function, suppression by morphine and habituation. J Neurophysiol 1993;70(2):630–9.
13. Hargreaves K, Dubner R, Brown F, Flores C, Joris J. A new and sensitive method for measuring thermal nociception in cutaneous hyperalgesia. Pain 1988;32(1):77–88.
14. Randall LO, Selitto JJ. A method for measurement of analgesic activity on inflamed tissue. Arch Int Pharmacodyn Ther 1957;111(4):409–19.
15. Galbraith JA, Mrosko BJ, Myers RR. A system to measure thermal nociception. J Neurosci Methods 1993;49(1–2):63–8.
16. Ankier SI. New hot plate tests to quantify antinociceptive and narcotic antagonist activities. Eur J Pharmacol 1974;27(1):1–4.
17. Van Ree JM, Leys A. Behavioral effects of morphine and phencyclidine in rats: the influence of repeated testing before and after single treatment. Eur J Pharmacol 1985;113(3): 353–62.
18. Bardo MT, Hughes RA. Exposure to a non-functional hot plate as a factor in the assessment of morphine-induced analgesia and analgesic tolerance in rats. Pharmacol Biochem Behav 1979;10(4):481–5.
19. Gamble GD, Milne RJ. Repeated exposure to sham testing procedures reduces reflex withdrawal and hot-plate latencies: attenuation of tonic descending inhibition? Neurosci Lett 1989;96(3):312–7.
20. Attal N, Jazat F, Kayser V, Guilbaud G. Further evidence for 'pain-related' behaviours in a model of unilateral peripheral mononeuropathy. Pain 1990;41(2):235–51.
21. Bennett GJ, Xie YK. A peripheral mononeuropathy in rat that produces disorders of pain sensation like those seen in man. Pain 1988;33(1):87–107.
22. Dixon WJ. Efficient analysis of experimental observations. Annu Rev Pharmacol Toxicol 1980;20:441–62.
23. Chaplan SR, Bach FW, Pogrel JW, Chung JM, Yaksh TL. Quantitative assessment of tactile allodynia in the rat paw. J Neurosci Methods 1994;53(1):55–63.
24. Otsuki T, Nakahama H, Niizuma H, Suzuki J. Evaluation of the analgesic effects of capsaicin using a new rat model for tonic pain. Brain Res 1986;365(2):235–40.
25. Gegout-Pottie P, Philippe L, Simonin MA, et al. Biotelemetry: an original approach to experimental models of inflammation. Inflamm Res 1999;48(8):417–24.
26. Combe R, Bramwell S, Field MJ. The mono-sodium iodoacetate model of osteoarthritis: a model of chronic nociceptive pain in rats? Neurosci Lett 2004;370(2–3):236–40.
27. Sluka KA, Milton MA, Willis WD, Westlund KN. Differential roles of neurokinin 1 and neurokinin 2 receptors in the development and maintenance of heat hyperalgesia induced by acute inflammation. Br J Pharmacol 1997;120(7):1263–73.
28. Vrinten DH, Hamers FF. 'CatWalk' automated quantitative gait analysis as a novel method to assess mechanical allodynia in the rat; a comparison with von Frey testing. Pain 2003;102(1–2):203–9.
29. Tonussi CR, Ferreira SH. Rat knee-joint carrageenin incapacitation test: an objective screen for central and peripheral analgesics. Pain 1992;48(3):421–7.
30. Sluka KA, Westlund KN. Behavioral and immunohistochemical changes in an experimental arthritis model in rats. Pain 1993;55(3):367–77.
31. Gauldie SD, McQueen DS, Clarke CJ, Chessell IP. A robust model of adjuvant-induced chronic unilateral arthritis in two mouse strains. J Neurosci Methods 2004; 139(2):281–91.
32. Skyba DA, Radhakrishnan R, Sluka KA. Characterization of a method for measuring primary hyperalgesia of deep somatic tissue. J Pain 2005;6(1):41–7.
33. Yu YC, Koo ST, Kim CH, Lyu Y, Grady JJ, Chung JM. Two variables that can be used as pain indices in experimental animal models of arthritis. J Neurosci Methods 2002;115(1): 107–13.

34. Han JS, Bird GC, Li W, Jones J, Neugebauer V. Computerized analysis of audible and ultrasonic vocalizations of rats as a standardized measure of pain-related behavior. J Neurosci Methods 2005;141(2):261–9.
35. Siegmund E, Cadmus R, Lu G. A method for evaluating both non-narcotic and narcotic analgesics. Proc Soc Exp Biol Med 1957; 95(4):729–31.
36. Hammond D. Inference of pain and its modulation from simple behaviors. In: Chapman C, Loeser J, eds. Issues in Pain Measurement: Advances in Pain Research and Therapy. New York: Raven Press; 1989;12:69–91.
37. Vyklicky L. Techniques for the study of pain in animals. In: Bonica J, Liebeskind J, Albe-Fessard D, eds. Advances in Pain Research and Therapy. New York: Raven Press; 1979;3:727–45.
38. Harada T, Takahashi H, Kaya H, Inoki R. A test for analgesics as an indicator of locomotor activity in writhing mice. Arch Int Pharmacodyn Ther 1979;242(2):273–84.
39. Dubuisson D, Dennis SG. The formalin test: a quantitative study of the analgesic effects of morphine, meperidine, and brain stem stimulation in rats and cats. Pain 1977;4(2):161–74.
40. Hunskaar S, Fasmer OB, Hole K. Formalin test in mice, a useful technique for evaluating mild analgesics. J Neurosci Methods 1985; 14(1):69–76.
41. Porro CA, Cavazzuti M. Spatial and temporal aspects of spinal cord and brainstem activation in the formalin pain model. Prog Neurobiol 1993;41(5):565–607.
42. Coderre TJ, Fundytus ME, McKenna JE, Dalal S, Melzack R. The formalin test: a validation of the weighted-scores method of behavioural pain rating. Pain 1993;54(1): 43–50.
43. Choi Y, Yoon YW, Na HS, Kim SH, Chung JM. Behavioral signs of ongoing pain and cold allodynia in a rat model of neuropathic pain. Pain 1994;59(3):369–76.
44. Wall PD, Devor M, Inbal R, et al. Autotomy following peripheral nerve lesions: experimental anaesthesia dolorosa. Pain 1979;7(2):103–11.
45. Hansson P. Difficulties in stratifying neuropathic pain by mechanisms. Eur J Pain 2003;7(4):353–7.
46. Mogil JS. The genetic mediation of individual differences in sensitivity to pain and its inhibition. Proc Natl Acad Sci U S A 1999;96(14): 7744–51.
47. Cizza G, Sternberg EM. The role of the hypothalamic-pituitary-adrenal axis in susceptibility to autoimmune/inflammatory disease. Immunomethods 1994;5(1):73–8.
48. West WL, Yeomans DC, Proudfit HK. The function of noradrenergic neurons in mediating antinociception induced by electrical stimulation of the locus coeruleus in two different sources of Sprague-Dawley rats. Brain Res 1993;626(1–2):127–35.
49. Mogil JS, Kest B, Sadowski B, Belknap JK. Differential genetic mediation of sensitivity to morphine in genetic models of opiate antinociception: influence of nociceptive assay. J Pharmacol Exp Ther 1996;276(2): 532–44.
50. Watling KJ, Guard S, Boyle SJ, McKnight AT, Woodruff GN. Species variants of tachykinin receptor types. Biochem Soc Trans 1994; 22(1):118–22.
51. Chapman CR, Nakamura Y. A passion of the soul: an introduction to pain for consciousness researchers. Conscious Cogn 1999;8(4):391–422.
52. Pasero C, Paice JA, McCaffery M. Basic mechanisms underlying the causes and effects of pain. In: McCaffery M, Pasero C, eds. Pain Clinical Manual. 2nd ed. St. Louis: Mosby Inc; 1999:15–34.
53. Byers M, Bonica JJ. Peripheral pain mechanisms and nociceptor plasticity. In: Loeser JD, Butler SH, Chapman CR, eds. Bonica's Management of Pain 3rd ed. Baltimore: Lippincott Williams & Wilkins; 2001:26–72.
54. Coda BA, Bonica JJ. General considerations of acute pain. In: Loeser JD, Butler SH, Chapman CR, eds. Bonica's Management of Pain. 3rd ed. Baltimore: Lippincott Williams & Wilkins; 2001:222–40.
55. Gebhart GF, Ness TJ. Central mechanisms of visceral pain. Can J Physiol Pharmacol 1991;69(5):627–34.
56. Portenoy RK. Mechanisms of clinical pain. Observations and speculations. Neurol Clin 1989;7(2):205–30.
57. Backonja MM. Painful Neuropathies. In: Loeser JD, Butler SH, Chapman CR, Turk DC, eds. Bonica's Management of Pain. 3rd ed. Baltimore: Lippincott Williams & Wilkins; 2001:371–87.
58. Galer BS, Schwartz L, Allen RJ. Complex regional pain syndromes—type I: reflex sympathetic dystrophy, and type II: causalgia. In: Loeser JD, Butler SH, Chapman CR, eds. Bonica's Management of Pain. 3rd ed. Baltimore: Lippincott Williams & Wilkins; 2001:388–411.
59. Tasker RR. Central pain states. In: Loeser JD, Butler SH, Chapman CR, eds. Bonica's Management of Pain. 3rd ed. Baltimore: Lippincott Williams & Wilkins; 2001:433–57.

Chapter 2

Animal Models of Inflammatory Pain

Rui-Xin Zhang and Ke Ren

Abstract

Animal models of inflammatory pain have been widely used to study the mechanisms of tissue injury-induced persistent pain. A variety of inflammatory agents or irritants, including complete Freund's adjuvant, carrageenan, zymosan, mustard oil, formalin, capsaicin, bee venom, acidic saline, lipopolysaccharide, inflammatory cytokines, and sodium urate crystals, have been used to produce tissue injury and hyperalgesia in such structures as cutaneous/subcutaneous tissues, joints, and muscles. Additionally, models of pain hypersensitivity have also been established with injuries produced by burning, freezing, and ultra irradiation. Although these models do not simulate every aspect of chronic pain, they do model key features of human inflammatory pain. Studies in animals give insight into certain aspects of human pain conditions and lead to improved pain management for patients.

1. Introduction

Pain perception is more complex in humans than in animals since human pain perception encompasses psychosocial, cultural, developmental, and environmental variables. However, human and animal pain perceptions show parallels, and animal models partially mimic the persistent pain encountered in the clinic. In the last two decades animal models of inflammatory pain have been widely used to study the mechanisms of tissue injury-induced persistent pain. Although none of the existing models can simulate all symptoms of inflammatory pain, studies in animals give insight into certain aspects of human pain conditions and lead to better pain management for patients. In the following paragraphs, commonly used inflammatory pain animal models will be summarized. Interested readers may consult more comprehensive reviews for further details (1, 2).

Chao Ma and Jun-Ming Zhang (eds.), *Animal Models of Pain*, Neuromethods, vol. 49,
DOI 10.1007/978-1-60761-880-5_2, © Springer Science+Business Media, LLC 2011

2. Inflammatory Models of Persistent Pain

Animal models of tissue injury and inflammatory hyperalgesia can be induced by a number of inflammatory agents in a variety of structures, including cutaneous and subcutaneous, joint, and muscle tissues.

2.1. Cutaneous and Subcutaneous Models of Inflammatory Pain

2.1.1. CFA Model

A *Mycobacterium butyricum* oil suspension was initially used to inoculate the tail base of the rat to induce adjuvant arthritis and persistent pain (3). Since polyarthritis develops after the inoculation along with a state of generalized illness, most pain researchers have discontinued the use of this model. However, the injection of complete Freund's adjuvant (CFA, composed of inactivated and dried *Mycobacterium* and adjuvant) into the footpad produces localized inflammation and persistent pain (4, 5). After a CFA injection into the footpad, cutaneous inflammation appears in minutes to hours and peaks within 5–8 h.

CFA produces dose-dependent inflammatory responses, and 30–200 μg of *Mycobacterium butyricum* suspended in oil/saline (1:1) yield significant edema and thermal hyperalgesia in the injected hind paw (6) (Fig. 1). The edema peaks around 24 h after the injection. The hyperalgesia and allodynia peak around 5 h after injection and persist for approximately 1–2 weeks (7).

CFA-induced hyperalgesia and allodynia in rats are consistent with those seen in humans receiving inadvertent injections of

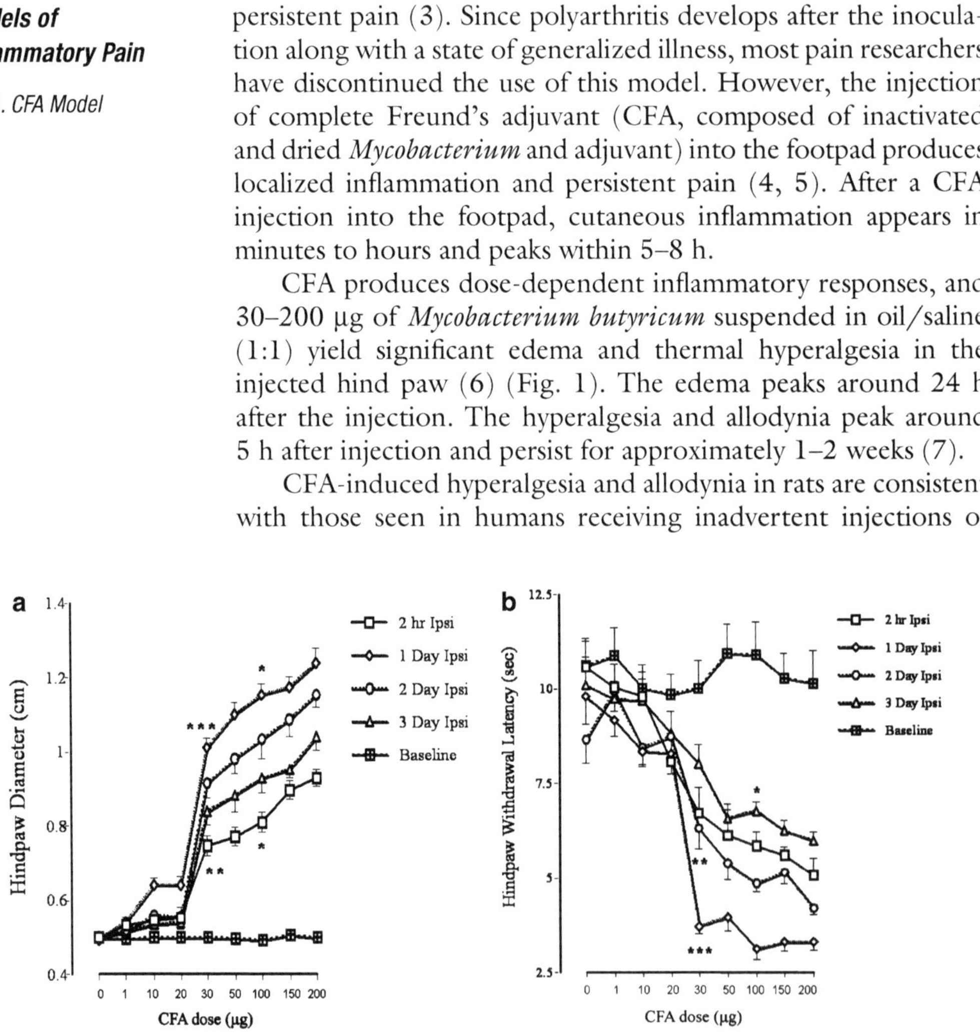

Fig. 1. Inflammation and hyperalgesia produced by intraplantar injection of complete Freund's adjuvant in rats. (**a**) Edema of the rat hind paw after injection of different doses of CFA, determined by measuring the dorsal-ventral thickness of the injected hindpaw. *$P<0.05$ compared to 200 μg injected rats; **$P<0.01$ compared to 20 μg injected rats; ***$P<0.001$ compared to 20 μg injected rats. (**b**) Changes in hind paw withdrawal latency to a noxious thermal stimulus at different time points (2 h to 3 days) after injection of different doses of CFA into the hindpaw. *$P<0.05$ compared to 200 μg injected rats; **$P<0.01$ compared to 20 μg injected rats; ***$P<0.001$ compared to 20 μg injected rats [reproduced with permission from Chinese Journal of Neuroanatomy (1999, 15:19–26), Chinese Society of Anatomical Science].

CFA (8). The physiological and biochemical effects of CFA are limited to the affected limb (5) and there are no signs of immune response or systemic disease. It has been shown that rats with CFA-induced inflammation exhibit minimal reductions in weight and show normal grooming behavior (5). Exploratory motor behavior is normal, and no significant alterations occur in an open field locomotion test (5).

Studies with three strains of rats, Lewis (LEW), Fischer 344 (FIS) and Sprague–Dawley (SD), demonstrate that, according to the difference scores computed by subtracting paw withdrawal latency (PWL) of the contralateral paw from that of the injected paw, F344 rats show significantly greater thermal hyperalgesia than do SD and LEW rats, both of which exhibit similar but relatively less intense hyperalgesia (9).

2.1.2. Carrageenan Model

An intraplantar injection of carrageenan is also widely used to produce a model of localized inflammatory pain. When 0.5 mg of carrageenan is injected, edema develops, mainly in two phases: the first 30 min after the injection, the second beginning at the end of the first hour and lasting until the third hour after injection. The edema peaks 3–5 h after injection (10, 11). When 6 mg of carrageenan is injected, edema peaks on day 3 (5) and thermal hyperalgesia peaks around 4 h after injection and lasts for at least 96 h (5). Studies with FIS, LEW, and SD rats demonstrate that LEW rats showed the least, and FIS rats the greatest, thermal hyperalgesia after intraplantar administration of 3.5 mg of carrageenan (12).

CFA and carrageenan are also injected into the facial area to study orofacial pain (13–15). A CFA injection into the perioral (PO) skin results in orofacial thermal hyperalgesia and mechanical allodynia that peak between 4 and 24 h and persist for at least 2 weeks (15). Facial carrageenan injection in mice causes increased responses to facial stimulation with a von Frey hair (1 g force) 8 h, 1 day, and 3 days after injection (13).

2.1.3. Mustard Oil and Zymosan Models

Other inflammatory agents such as mustard oil, a small fiber irritant, and zymosan, a glucan from the cell walls of yeast, have been used to produce behavioral hyperalgesia. Mustard oil elicits inflammatory pain by activating transient receptor potential cation channel, subfamily A, member 1 (TRPA1), an excitatory ion channel of primary afferent nociceptors (16). Topical application of mustard oil to the ear induces dose-dependent increases in plasma extravasation and ear thickness, which peak approximately 30 min after application (17). Application of mustard oil to rat paw skin induces plasma protein extravasation and slight edema, a 7–8% increase in paw volume (18). Topical application of mustard oil (20 μl, 100%) to the lateral surface of the left hind leg induces immediate agitation with frequent biting and vocalizations. This response lasts approximately 5–7 min. Mustard oil also significantly facilitates a tail-flick reflex that appears 5 min after application and lasts up to 60 min, peaking 20 min after application (19).

Intraplantar injection of zymosan (0.31–6.25 mg) produces persistent dose- and time-dependent mechanical and thermal hyperalgesia. Edema is greatest at dosages ≥2.5 mg and peaks 30-min postinjection irrespective of dosage. Mechanical hyperalgesia appears at dosages ≥1.25 mg and reaches its maximum 4 h after application at a dosage of 5 mg. Thermal hyperalgesia is biphasic and dose-dependent. An early-phase peak occurs at 30 min at dosages ≥2.5 mg; this is not apparent at lower dosages. A late-phase peak occurs at 4 h at dosages of ≥0.0625 mg. Higher dosages (5 and 6.25 mg) also cause spontaneous pain, sometimes characterized by occasional flicking of the hind paw but more commonly by elevation of the paw for extended periods of time (20).

2.1.4. Formalin Model

The formalin test is a popular model for studying pain mechanisms under prolonged nociception. Formalin is injected beneath the footpad of a rat, mouse, or cat and produces two phases of nocifensive behavior, characterized by licking and flinching of the paw, that are separated by a short period of quiescence (21, 22). The first or acute phase occurs typically in the first 5 min; the second starts from 15 min and lasts about 40–60 min after injection. It is generally agreed that the first phase is due to the direct activation of both low-threshold mechanoreceptive and nociceptive primary afferent fibers (23). There has been disagreement about the underlying mechanisms of the second phase. Early studies suggested that the second phase resulted from an increase in the excitability of dorsal horn neurons. More recently, it has been demonstrated that ongoing activity of primary afferent fibers is necessary for the development of the second phase (23–26). In regard to the period of quiescence, some evidence supports the idea of an absence of activity, other evidence implicates an active inhibitory mechanism (27).

Formalin-induced pain is measured by combining scores of favoring, lifting, licking, and flinching/shaking of the injured paw. Power analysis indicates that using a moderate dosage (1.5%, 0.05 ml) of formalin and a combined pain score gives the greatest power to detect pain differences (22). Further, combining the formalin model with the place-conditional paradigm demonstrates that, when compared with a distinct environmental context, a hind paw injection of formalin induces conditioned place avoidance, which reflects a negative affective state (28).

To study orofacial pain, formalin is subcutaneously injected into the rat upper lip or lateral face and generates similar biphasic behavioral responses (face rubbing), an early and short-lasting first phase followed, after a quiescent period, by a second prolonged (tonic) phase. The orofacial formalin test can be used to produce and quantify nociception in the trigeminal region of the rat (29–31). Study with different formalin concentrations indicates that the second phase response to formalin only occurs with dosages of 1.5% and higher (31).

2.1.5. Bee Venom Model

A subcutaneous injection of bee venom (0.2-mg lyophilized whole venom in 0.1-ml saline) into the hind paw produces persistent nociceptive responses (flinching and lifting/licking the injected paw) for 1–2 h, followed by a 72–96 h period of mechanical allodynia and thermal hyperalgesia accompanied by edema and redness of the injected paw. It also produces thermal hyperalgesia, but not mechanical allodynia, on the contralateral hind paw although with less amplitude than that of the injected paw (32).

2.1.6. Capsaicin Model

Capsaicin, the pungent component of cayenne pepper that activates transient receptor potential vanilloid type 1 (TRPV1), a heat-sensitive cation channel on nociceptor terminals, has been used in humans and animals to study neurogenic inflammation and hyperalgesia. Intradermal injection of capsaicin results in flare reaction, allodynia, and hyperalgesia, the areas of which extend beyond the injection site. Visual observation of flare response reveals that the area of visual flare is significantly smaller than the area of hyperalgesia to stroking stimuli and that the latter is significantly smaller than that for punctate stimuli. The heat hyperalgesia (thermode maintained at 38°C) area is the smallest (33, 34). Thermographic detection of the flare response shows that the thermographic area is larger than the area of visual flare and coincides with the area of mechanical (nylon monofilament, 1.02-mm diameter exerting a bending force of 2.02 N) and heat hyperalgesia (from a 1-cm^2 Peltier thermode maintained at 47°C) (35). Regarding the temporal pattern of flare, visual flare reaches its maximum within 3–5 min (33). Laser-Doppler flowmetry also shows that blood flow reaches maximum 5 min after the injection and then decreases (34). A thermographic device shows that the flare response starts as early as a few seconds after the capsaicin injection (35). The area of hyperalgesia to stroking stimuli appears immediately after injection, peaks within 15 min and then gradually decreases over 1–6 h. The area of hyperalgesia to punctate stimulation is immediately present after injection, grows to a maximum within 15–30 min, decreases gradually, and disappears at about 21 h. The area of the heat hyperalgesia reaches maximum 30 min after injection, gradually decreases, and disappears about 1.5 h after injection (33). Capsaicin (0.1, 1, 10, and 100 μg) produces dose-dependent increases in spontaneous pain, area and intensity of mechanical allodynia, area and intensity of pinprick hyperalgesia, and flare area (36).

It should be noted that capsaicin may produce differential responses in different areas. For instance, peak pain intensity and duration are greater in the forehead than in the forearm, while areas of visible flare and pinprick hyperalgesia are significantly larger in the forearm than in the forehead (37).

This neurogenic model of inflammation has been used in monkeys to study changes in nociceptor activity and changes in the responses of spinal dorsal horn neurons (33, 38). It has recently been adapted for behavioral studies in rats (39). Intraplantar injection of capsaicin evokes nocifensive behavior characterized by lifting and guarding of the injected paw that lasts for about 3 min. Capsaicin dose-dependently produces thermal and mechanical hyperalgesia. Thermal hyperalgesia to heat lasts up to 45 min, whereas mechanical hyperalgesia persists up to 4 h.

To study trigeminal pain, subcutaneous injection of different dosages of capsaicin into the vibrissa pad produces an immediate rubbing–scratching of the injected area. This behavior is performed with the ipsilateral forepaw, often accompanied by the contralateral forepaw. The rubbing–scratching response reaches its maximum during the 12- to 18-min interval and subsides about 42 min after capsaicin injection. Morphine dose-dependently reduces the capsaicin-induced rubbing–scratching (40, 41).

Further, capsaicin-sensitive primary afferents play different roles in chemical irritant-induced spontaneous nociception, hyperalgesia, and inflammatory responses (42). Local pretreatment with capsaicin significantly inhibits the two phases of formalin-induced persistent spontaneous nociception, while it only inhibits the late phase (tonic nociception; 11–60 min), not the early phase (acute nociception; 0–10 min), of spontaneous nociception in the bee venom test. Although capsaicin pretreatment prevents thermal hyperalgesia in the bee venom, carrageenan, and CFA models, it only prevents mechanical allodynia in the bee venom and carrageenan models, not the CFA model. Regarding inflammatory response, it significantly inhibits bee venom-elicited paw edema but not carrageenan-, CFA-, or formalin-elicited paw edema.

Rank order of the duration of the inflammatory hyperalgesia produced by these agents, is, from longest to shortest, CFA > bee venom > carrageenan > zymosan > formalin > mustard oil. See Table 1 for a comparison of onsets and durations.

2.2. Models of Joint Inflammation and Hyperalgesia

2.2.1. CFA-Induced Joint Inflammation

Adjuvant arthritis is induced in rats for a laboratory animal model of chronic pain that mimics that of human rheumatic disease. A CFA injection into the base of the rat's tail causes polyarthritis (43). Hypersensitivity of multiple joints occurs after 10 days and lasts up to 3 weeks. The paws and tails of arthritic rats show lower thresholds in response to noxious pressure (hyperalgesia), higher thresholds in response to noxious heat (hypoalgesia), and no change in response to noxious electrical stimulation (44).

Table 1
Comparison of cutaneous/subcutaneous inflammatory pain models[a]

Chemical	Hyperalgesia	Allodynia	Time of onset	Duration
CFA	Yes	Yes	2–6 h	1–2 weeks
Carrageenan	Yes	Yes	1 h	24 h
Mustard oil	Yes	Yes	5 min	<1 h
Zymosan	Yes	Yes	30 min	24 h
Formalin phase I	N/A[b]	N/A	<1 min	5–10 min
Formalin phase II	N/A	N/A	10 min	1 h
Bee venom	Yes	Yes	1 min	96 h
Capsaicin	Yes	Yes	1 min	<1 h

[a]Modified and reprinted with permission from the ILAR Journal, 40(3),1999, Institute for Laboratory Animal Research, The National Academics, 500 Fifth Street NW, Washington, DC 20001 (http://www.national-academies.org/Ilar)

[b]Not applicable

Pain is also inferred from scratching behaviors, reduced motor activity, weight loss, vocalization when the affected limbs are pinched, and a reduction in these behaviors following the administration of opioids. It should be noted that this is a systemic disease that includes skin lesions, destruction of bone and cartilage, impairment of liver function, and lymphadenopathy, which leads to ethical concerns (45). Moreover, the systemic lesions make it difficult to differentiate pain behavior from generalized malaise and debilitation. The likely presence of central nervous system changes associated with the alterations in immune function also call into question the use of this model as a correlation of pain behavior to neural activity and neurochemical alterations. Polyarthritis is also induced by type II collagen emulsified in Freund's incomplete adjuvant and injected into the base of the rat's tail. Tail-flick latency significantly decreases 1 week after inoculation, peaks at 3 weeks, and lasts for at least 5 weeks (46). Like CFA-induced polyarthritis, collagen-induced polyarthritis raises ethical concerns.

An alternative to the polyarthritic rat for prolonged studies, intra-knee joint injection of 250-mg suspension of heat-killed *Mycobacterium butyricum* in peanut oil and saline (1:1), causes ipsilateral thermal hyperalgesia of the hind paw 1 day after injection, peaks on day 3, and remains at that level until day 14. By the 21st day, experimental animals recover from the hyperalgesia (47). Knee joint withdrawal threshold measurement, in which a gradually increasing squeeze is applied across the joint, shows that the average ipsilateral limb withdrawal threshold (LWT) significantly decreases by 57%, from 790 ± 39 to 316 ± 45 g, 1 day after the CFA injection. This decrease lasts for a full 28 days, the period studied. An incapacity test shows that the ratio of weight distribution between ipsilateral and contralateral limbs significantly decreases from 0.96 ± 0.03 to 0.29 ± 0.05. This significant decrease lasts up to day 28 (48).

Injection (0.05 ml) of 300 μg *Mycobacterium butyricum* into the tibio-tarsal joint also produces monoarthritis. As revealed by clinical observations and X-ray examinations, the arthritis is limited anatomically, pronounced, prolonged, and stable from weeks 2 through 6 postinjection. The affected limb shows a marked increase in sensitivity to paw pressure. Animals gain weight, remain active, and evince little systemic disturbance in contrast to polyarthritic rats (49).

CFA also produces significant thermal hyperalgesia and mechanical allodynia following its injection into the temporomandibular joint. Thermal hyperalgesia develops at 5 h, peaks at 24 h, and lasts 2 weeks. Mechanical allodynia starts at 2 h, peaks between 4 and 24 h and persists for at least 2 weeks after the injection (15).

2.2.2. Kaolin and Carrageenan-Induced Joint Inflammation

Another arthritis animal model is induced by injecting kaolin and carrageenan (3 mg/3 mg) into the knee joint. In rats, decrease in PWL occurs ipsilaterally to the inflamed knee as early as 4 h after an injection of the two agents and lasts about 24 h. The circumference of the ipsilateral knee joint is significantly larger than baseline between 4 and 24 h. The rats also show spontaneous pain, indicated by decrease of weight bearing by the injected limb (50, 51). Recently, the elevated plus maze test has demonstrated that arthritic rats show amygdala-involved anxiety-like behavior, evidenced by a decreased preference for the open arms (52). Intra-articular injection of carrageenan and kaolin in cats causes guarding of the leg and avoidance of movement or weight bearing. These symptoms begin ~2 h after the injection, are fully developed after 4 h, and last at least 15 h.

Recording of saphenous nerve filament activity consistently demonstrates that almost all filament units of inflamed joints have low thresholds to passive movement of the knee joint compared to units of normal joints. The number of receptive fields per unit is significantly greater than that seen in normal joints (53). In rats, cats, and monkeys, somatosensory neurons of the spinal cord become hyperexcitable to mechanical stimuli during the development of experimental arthritis (54). Changes in joint receptors and spinal dorsal horn neuronal activity begin as soon as 1–2 h following injection and build for several hours. Noticeably, the magnitude of hyperalgesia in this model is relatively low compared to that of the hind paw inflammation model. It should be kept in mind that the test site, the paw, in the joint inflammation model is remote from the injury site. What is measured by PWL in the knee joint inflammation model is, likely, only secondary hyperalgesia.

2.2.3. Carrageenan-Induced Joint Inflammation

Compared to the short-lasting effects of the kaolin and carrageenan combination, carrageenan alone produces a long-lasting effect. An intra-articular injection of 3 mg of carrageenan significantly decreases ipsilateral and contralateral PWL to heat; the decrease occurs at 4 h and lasts 6 weeks. Carrageenan at 0.3 and 1 mg produces only ipsilateral effects that are shorter-lasting: 24 h for 0.3 mg and up to 3 weeks for 1 mg. Intra-articular injection of 3 mg of carrageenan also produces significant decrease of the mechanical withdrawal threshold ipsilaterally (between 3 and 7 weeks) and contralaterally (between 3 and 6 weeks). Carrageenan at 1 mg induces a significant ipsilateral decrease between 4 h and 1 week, but 0.3 mg has no effect (55).

2.2.4. Formalin-Induced Joint Inflammation

Formalin (0.5, 3, and 5%) injected into the knee joint of rats induces dose-dependent nocifensive responses. The nociception consists of two phases (from 0 to 5 min and from 10 to 60 min) of intense guarding behavior on the affected limb with an intervening period of quiescence (from 5 to 10 min). Morphine (4 mg/kg, subcutaneously) pretreatment reduces the guarding behavior in both nocifensive phases (56). A formalin (0.5, 2.5, and 5%) injection into the temporomandibular joint (TMJ) region induces a dose-dependent phase of orofacial rubbing and one of flinching, alternately displayed. The rubbing responses start earlier, peak 18 min postinjection, and then decrease; the flinching responses start later, peak 27 min postinjection, and last up to 36 min. A significant correlation exists between formalin concentration and rubbing and flinching responses. The magnitude of these responses reaches its maximum at a concentration of 2.5% (57).

2.2.5. Joint Inflammation Induced by Other Irritants

Other models of arthritis have been developed using sodium urate crystals, which are injected into the ankle joint of a rat or cat (45, 58). The arthritis is fully developed within 24 h. These animals tend to place less weight on the treated hind limb and exhibit guarding movements of the limb. In the rat, touch, pressure and thermal stimuli applied to the affected paw result in a decrease in responsiveness, presumably due to pain associated with the movement. There are no signs of systemic disease in the urate arthritis model other than the joint pathology secondary to tissue edema and the infiltration of polymorphonuclear leukocytes (45).

Capsaicin (0.2%, 50 μl) injected into the lateral aspect of the left ankle joint results in a decreased mechanical withdrawal threshold 2 h after injection; this is maintained through a 4-h period (59). Capsaicin-sensitive primary afferents play different roles in a variety of joint inflammation models, as they do in cutaneous/subcutaneous pain models. Pretreating the joint with 1% capsaicin (about 1 week before injection) significantly reduces the inflammatory response to carrageenan and urate but not to formalin (60). Onsets and pain durations produced by joint inflammation are compared in Table 2.

Table 2
Comparison of joint inflammatory pain models

Chemical	Hyperalgesia	Allodynia	Time of onset	Duration
CFA	Yes	Yes	1 day	3–4 weeks
Carrageenan	Yes	Yes	4 h	6 weeks
Formalin phase I	N/A[a]	N/A	<1 min	5 min
Formalin phase II	N/A	N/A	10 min	1 h
Capsaicin	N/A	Yes	2 h	4 h

[a]Not applicable

2.3. Models of Muscle Inflammation and Hyperalgesia

2.3.1. Acidic Saline-Induced Muscle Inflammation and Hyperalgesia

The bulk of available knowledge about pain mechanisms is derived from studies on cutaneous pain. However, the existing subjective differences between muscle and skin pain (e.g., muscle pain is poorly localized and shows referral) suggest that muscle pain has distinct characteristics. Models have been developed to examine mechanisms underlying the development and maintenance of chronic muscle pain.

A decrease in tissue pH has been observed in response to inflammation, hematomas, and isometric exercise. Decreasing pH increases activity of nociceptors and produces a painful response in humans (61). Using an in vitro nerve-skin preparation, continuous infusion of low pH (5.2–6.9) saline increases discharges of C-polymodal primary afferent fibers without adaptation (62). Repeated injection of low pH saline into the gastrocnemius muscle of rats produces long-lasting, widespread mechanical hyperalgesia without motor deficits or significant tissue damage.

Following the first unilateral intra-gastrocnemius muscle injection of pH 4.0, 5.0, or 6.0 saline, the mechanical withdrawal threshold of the ipsilateral paw dose-dependently decreases 4 h, and returns to baseline 24 h, after injection. After a second unilateral injection of low pH saline on day 5, the mechanical withdrawal threshold dose-dependently decreases in both ipsilateral and contralateral hind paws. These bilateral decreases are greatest for pH 4.0 saline, persisting for 4 weeks after the second injection. Inter-injection intervals of 2 and 5 days (pH 4.0 saline) produce equivalent and significant bilateral decreases in mechanical withdrawal threshold. A 10-day inter-injection interval does not produce persistent mechanical hyperalgesia. This suggests that there is a critical window in which re-injury to muscle tissue results in exaggerated, more persistent hyperalgesia. Intra-gastrocnemius muscle lidocaine 24 h after the second injection of pH 4.0 saline increases the ipsilateral mechanical withdrawal threshold during the first 10–15 min after injection. However, lidocaine has no significant effect on the decreased contralateral withdrawal threshold.

Withdrawal latencies to radiant heat average approximately 10 s at baseline and are no different than controls after either the first or second injection of low pH saline. After both the first and the second injection of low pH saline, rats show no limb guarding, have equal weight bearing and normal gait patterns, and their ability to perform the treadmill test is unchanged (63). Interestingly, in sharp contrast to most persistent pain models intramuscular acidic saline-induced allodynia does not involve spinal glial activation or inflammatory cytokine interleukin-1 (64).

2.3.2. Carrageenan-Induced Muscle Inflammation and Hyperalgesia

Intramuscular application of carrageenan sensitizes group III and IV muscle afferents, including nociceptors, lowering their thresholds to mechanical activation and increasing their background activity. Furthermore, carrageenan induces local inflammation when injected into the muscle, as evidenced by the accumulation of leukocytes that begins 2 h postinjection and continues for the next 8 h (65).

Injection of carrageenan (0.5–6 mg/triceps) into the bilateral triceps muscles produces dose-dependent reduction in forelimb grip force that peaks 24 h, and returns to the control level 48 h, postinjection. Capsaicin (50 mg/kg i.p.) admi nistration to rats on the second day of life reduces carrageenan-evoked hyperalgesia by about 45%, indicating that the muscle hyperalgesia induced by carrageenan is mediated, in part, by capsaicin-sensitive afferent fibers (66).

Unilateral intra-gastrocnemius muscle injection of carrageenan also dose-dependently produces thermal and mechanical hyperalgesia. Carrageenan at 3 mg significantly decreases ipsilateral PWL to heat that occurs within 4 h and lasts 8 weeks. Contralateral PWL also decreases by the end of the first week and last 8 weeks. Carrageenan at 0.3 and 1 mg produces no significant effect on PWL to noxious thermal stimulus (55). However, a 3-mg intramuscle injection of carrageenan significantly decreases the ipsilateral mechanical withdrawal threshold between 4 h and 6 weeks and the contralateral threshold between weeks 3 and 6. The ipsilateral mechanical withdrawal threshold significantly decreases between 4 and 8 h after a 1-mg intramuscle injection of carrageenan. Carrageenan at a dosage of 0.3 mg produces no changes in mechanical sensitivity (55). It seems that mechanical sensation is more sensitive to carrageenan than is thermal sensation.

2.3.3. Cytokine-Induced Muscle Inflammation and Hyperalgesia

An intra-gastrocnemius injection of tumor necrosis factor-alpha (TNF) significantly decreases mechanical withdrawal thresholds to muscle pressure in rats when measured with an algesimeter that exerts pressure on the gastrocnemius muscle. It also decreases forelimb grip strength as measured with a digital grip force meter. The hyperalgesia lasts at least 60 min (67). Similarly, an intra-gastrocnemius injection of another pro-inflammatory cytokine interleukin-6 significantly decreases the mechanical threshold, which returns to normal levels 5 days after the injection (68).

2.3.4. Mustard Oil, Formalin, and Hypertonic Saline

Mustard oil (30 µl, 20%), formalin (50 µl, 3%), or hypertonic saline (100 µl, 5%) injected in the mid-region of the masseter muscle elicits significant hind paw shaking behavior compared to an injection of vehicle. The peak and overall magnitude of this behavior present in a dose-dependent manner (mustard oil 1–20%) and is dose-dependently attenuated by the systemic administration of morphine sulfate (69). CFA injected into the bilateral masseter muscles of the rat significantly decreases the bite force on days 1, 2, and 3. The bite force gradually increases and then returns to baseline by day 14 (70). Onsets and durations of pain produced by muscle inflammation are compared in Table 3.

2.4. Other Inflammatory Pain Models

2.4.1. Burn Injury Model

Burn injury-induced pain model has been developed in rats. A mild focal thermal injury (52°C/30–45 s) to the rat heel induces primary thermal and mechanical hyperalgesia that lasts for 2 h and secondary mechanical hyperesthesia and tactile allodynia that last 3 and 2 h, respectively (71). After a third-degree burn injury induced by immersing the dorsal part of the right hind paw into a hot water bath (85°C/12 s), the rats display thermal hyperalgesia and mechanical allodynia that are evident by day 1, peak around days 5–7, and persist for at least 2 weeks (72). The dose–response curve of morphine shifts to the right in burn-injured rats as compared to sham rats on postinjury day 14 (72). This is in contrast to carrageenan-induced inflammatory pain, in which the dose–response curves for intrathecal mu- and delta-opioid receptor agonists shift to the left for inflamed hind paws compared to contralateral, uninflamed paws (73).

A burn injury human model has been reported. The injury is produced on the medial part of the nondominant crus, and pain thresholds decrease and pain responses increase in reaction to both thermal and mechanical stimuli within the burn area according to a visual analog scale (0–100). The burns also induce

Table 3
Comparison of muscle inflammatory pain models

Chemical	Hyperalgesia	Allodynia	Time of onset	Duration
Acidic saline	No	Yes	4 h	24 h (one inj.) or 4 weeks (two inj.)
Carrageenan	Yes	Yes	4 h	8 weeks (hyperalgesia) 6 weeks (allodynia)
Interleukin-6	N/A	Yes	30 min	3 h to 5 days

secondary mechanical and thermal hyperalgesia and increased pain response to mechanical and heat stimuli in the area of secondary hyperalgesia (74).

2.4.2. Freeze Injury Model

Recently, a freeze injury-induced pain model has been developed in humans. The end (1.8 cm^2) of a cylindrical copper bar previously cooled to −28°C is applied on the anterior glabrous part of the forearm to induce a first-degree burn injury. Several hours after the injury, sharply delimited erythema accompanied by localized hyperalgesia is observed. Tissue injury leads to primary mechanical hyperalgesia, limited to the area of injury, and to secondary mechanical hyperalgesia in the undamaged skin surrounding the injury. The hyperalgesia lasts at least 70 h and can be significantly relieved by systemic or topical ibuprofen. This is a useful tool for evaluating the efficacy and detecting the potential sites of action of analgesic agents such as nonsteroidal anti-inflammatory drugs in healthy human subjects (75).

2.4.3. Ultraviolet Irradiation Model

Acute cutaneous over-exposure to ultraviolet radiation (UVR) has been used to construct inflammatory pain models in both rats and humans. UVB (280–320 nm) irradiation of the hairy skin of rat hind paws dose-dependently produces erythema, thermal hyperalgesia, and mechanical allodynia (76). Erythema occurs on day 1 postirradiation, peaks on day 2, and returns to normal between days 4 and 7 depending on the UV dosage. Thermal hyperalgesia and mechanical allodynia appear on day 1, peak on day 2, and return to normal between days 7 and 14 for thermal sensation and days 4 and 14 for mechanical sensation, depending on the UV dosages. Systemic morphine produces dose-dependent and naloxone-sensitive reversal of the sensory changes.

In humans, UV irradiation also causes erythema and thermal hyperalgesia and mechanical allodynia (77). Solar-simulated radiation (SSR) at three minimal erythema dosages (MED) induces significant erythema 3 h postirradiation that peaks in 24 h and lasts at least 72 h. SSR dose-dependently produces significant thermal hyperalgesia 24 h postirradiation that lasts at least 72 h, and mechanical hyperalgesia in 6 h that peaks in 48 h and lasts at least 72 h.

2.4.4. Toxin-Induced Pain Models

Snakebites are a public health problem in Central and South America. It has been reported that an intraplantar injection (5–20 μg/paw) of Lys49 or Asp49 phospholipases A2 from *Bothrops asper* snake venom causes mechanical hyperalgesia, measured with the paw pressure test, which peaks in 1 h and normalizes 24 h after the injection (78).

In the aforementioned inflammatory pain models, the peripheral somatic nociceptors are stimulated and the nociceptive inputs are directly transmitted to the spinal cord or to trigeminal sensory nuclei. In contrast, intraperitoneally administered lipopolysaccharide (LPS; bacterial endotoxin) does not result in direct nociceptive input to the spinal cord dorsal horn but produces a long-lasting facilitation of the nociceptive tail-flick reflex (79). It has been shown that LPS-induced hyperalgesia is blocked by hepatic vagotomy and descending funiculus lesions but not by transection of the splanchnic nerve, the primary afferent for visceral pain (80). LPS-induced hyperalgesia lasts at least 60 min.

3. Conclusion

A variety of inflammatory pain animal models have been established by mimicking the injury of different tissues of skin, muscle, and joint. Although these models do not simulate every aspect of chronic pain symptoms, they do imitate some key features of human inflammatory pain. Through studies with a variety of persistent pain models, enormous progress is being made in the discovery of the cellular and molecular mechanisms responsible for the pathogenesis of pain.

Acknowledgment

We thank Dr. Lyn Lowry for her editorial support. The authors' work is supported by NIH Grants AT002605, NS060735, and DE11964.

References

1. Bouncier, M., Cavey, D., Kail, N., and Hensby, C. (1990) Experimental models in skin pharmacology, *Pharmacological Reviews 42*, 127–154.
2. Le Bars, D., Gozariu, M., and Cadden, S. W. (2001) Animal models of nociception, *Pharmacological Reviews 53*, 597–652.
3. Colpaert, F. C. (1987) Evidence that adjuvant arthritis in the rat is associated with chronic pain, *Pain 28*, 201–222.
4. Millan, M. J., Czlonkowski, A., Morris, B., Stein, C., Arendt, R., Huber, A., Hollt, V., and Herz, A. (1988) Inflammation of the hind limb as a model of unilateral, localized pain: influence on multiple opioid systems in the spinal cord of the rat, *Pain 35*, 299–312.
5. Iadarola, M. J., Brady, L. S., Draisci, G., and Dubner, R. (1988) Enhancement of dynorphin gene expression in spinal cord following experimental inflammation: stimulus specificity, behavioral parameters and opioid receptor binding, *Pain 35*, 313–326.
6. Zhang, R.-X., Chen, J. Y., Ruda, M. A., and Qiao, J.-T. (1999) Functional relationship between peripheral inflammation, thermal hyperalgesia, and enhanced dynorphin gene expression in spinal cord of rat: behavioral and mRNA studies, *Chinese Journal of Neuroanatomy 15*, 19–26.
7. Lao, L., Zhang, R.-X., Zhang, G., Wang, X., Berman, B. M., and Ren, K. (2004) A parametric study of electroacupuncture on persistent

hyperalgesia and Fos protein expression in rats, *Brain Research 1020*, 18–29.

8. Gould, H. J. III. (2000) Complete Freund's adjuvant-induced hyperalgesia: a human perception, *Pain 85*, 301–303.
9. Zhang, R.-X., Lao, L., Qiao, J.-T., and Ruda, M. A. (2003) Strain differences in pain sensitivity and expression of preprodynorphin mRNA in rats following peripheral inflammation, *Neuroscience Letters 353*, 213–216.
10. Vinegar, R., Schreiber, W., and Hugo, R. (1969) Biphasic development of carrageenin edema in rats, *Journal of Pharmacology and Experimental Therapeutics 166*, 96–103.
11. Winter, C. A., Risley, E. A., and Nuss, G. W. (1962) Carrageenin-induced edema in hind paw of the rat as an assay for antiiflammatory drugs, *Proceedings of the Society for Experimental Biology and Medicine 111*, 544–547.
12. Fecho, K., Nackley, A. G., Wu, Y., and Maixner, W. (2005) Basal and carrageenan-induced pain behavior in Sprague-Dawley, Lewis and Fischer rats, *Physiology & Behavior 85*, 177–186.
13. Yeo, J. F., Ong, W. Y., Ling, S. F., and Farooqui, A. A. (2004) Intracerebroventricular injection of phospholipases A2 inhibitors modulates allodynia after facial carrageenan injection in mice, *Pain 112*, 148–155.
14. Ng, C. H. and Ong, W. Y. (2002) Increased synaptosomal [3H] GABA uptake in the rat brainstem after facial carrageenan injections, *Pain 98*, 259–268.
15. Imbe, H., Iwata, K., Zhou, Q. Q., Zou, S., Dubner, R., and Ren, K. (2001) Orofacial deep and cutaneous tissue inflammation and trigeminal neuronal activation. Implications for persistent temporomandibular pain, *Cells Tissues Organs 169*, 238–247.
16. Bautista, D. M., Jordt, S. E., Nikai, T., Tsuruda, P. R., Read, A. J., Poblete, J., Yamoah, E. N., Basbaum, A. I., and Julius, D. (2006) TRPA1 mediates the inflammatory actions of environmental irritants and proalgesic agents, *Cell 124*, 1269–1282.
17. Inoue, H., Asaka, T., Nagata, N., and Koshihara, Y. (1997) Mechanism of mustard oil-induced skin inflammation in mice, *European Journal of Pharmacology 333*, 231–240.
18. Lippe, I. T., Stabentheiner, A., and Holzer, P. (1993) Participation of nitric oxide in the mustard oil-induced neurogenic inflammation of the rat paw skin, *European Journal of Pharmacology 232*, 113–120.
19. Urban, M. O., Jiang, M. C., and Gebhart, G. F. (1996) Participation of central descending nociceptive facilitatory systems in secondary hyperalgesia produced by mustard oil, *Brain Research 737*, 83–91.
20. Meller, S. T. and Gebhart, G. F. (1997) Intraplantar zymosan as a reliable, quantifiable model of thermal and mechanical hyperalgesia in the rat, *European Journal of Pain 1*, 43–52.
21. Dubuisson, D. and Dennis, S. G. (1977) The formalin test: a quantitative study of the analgesic effects of morphine, meperidine, and brain stem stimulation in rats and cats, *Pain 4*, 161–174.
22. Abbott, F. V., Franklin, K. B. J., and Westbrook, R. F. (1995) The formalin test: scoring properties of the first and second phases of the pain response in rats, *Pain 60*, 91–102.
23. Puig, S. and Sorkin, L. S. (1996) Formalin-evoked activity in identified primary afferent fibers: systemic lidocaine suppresses phase-2 activity, *Pain 64*, 345–355.
24. Pitcher, G. M. and Henry, J. L. (2002) Second phase of formalin-induced excitation of spinal dorsal horn neurons in spinalized rats is reversed by sciatic nerve block, *European Journal of Neuroscience 15*, 1509–1515.
25. Taylor, B. K., Peterson, M. A., and Basbaum, A. I. (1995) Persistent cardiovascular and behavioral nociceptive responses to subcutaneous formalin require peripheral nerve input, *Journal of Neuroscience 15*, 7575–7584.
26. Abbadie, C., Taylor, B. K., Peterson, M. A., and Basbaum, A. I. (1997) Differential contribution of the two phases of the formalin test to the pattern of c-fos expression in the rat spinal cord: studies with remifentanil and lidocaine, *Pain 69*, 101–110.
27. Henry, J. L., Yashpal, K., Pitcher, G. M., and Coderre, T. J. (1999) Physiological evidence that the 'interphase' in the formalin test is due to active inhibition, *Pain 82*, 57–63.
28. Johansen, J. P., Fields, H. L., and Manning, B. H. (2001) The affective component of pain in rodents: direct evidence for a contribution of the anterior cingulate cortex, *Proceedings of the National Academy of Sciences of the United States of America 98*, 8077–8082.
29. Raboisson, P. and Dallel, R. (2004) The orofacial formalin test, *Neuroscience and Biobehavioral Reviews 28*, 219–226.
30. Yeo, J. F. (2002) Does nitric oxide play a role in orofacial pain transmission? *Annals of the New York Academy of Sciences 962*, 151–160.
31. Clavelou, P., Dallel, R., Orliaguet, T., Woda, A., and Raboisson, P. (1995) The orofacial formalin test in rats: effects of different formalin concentrations, *Pain 62*, 295–301.

32. Chen, J., Luo, C., Li, H.-L., and Chen, H.-S. (1999) Primary hyperalgesia to mechanical and heat stimuli following subcutaneous bee venom injection into the plantar surface of hindpaw in the conscious rat: a comparative study with the formalin test, *Pain 83*, 67–76.
33. LaMotte, R. H., Shain, C. N., Simone, D. A., and Tsai, E. F. (1991) Neurogenic hyperalgesia: psychophysical studies of underlying mechanisms, *Journal of Neurophysiology 66*, 190–211.
34. Sumikura, H., Andersen, O. K., Drewes, A. M., and Arendt-Nielsen, L. (2003) Spatial and temporal profiles of flare and hyperalgesia after intradermal capsaicin, *Pain 105*, 285–291.
35. Serra, J., Campero, M., and Ochoa, J. (1998) Flare and hyperalgesia after intradermal capsaicin injection in human skin, *Journal of Neurophysiol 80*, 2801–2810.
36. Scanlon, G. C., Wallace, M. S., Ispirescu, J. S., and Schulteis, G. (2006) Intradermal capsaicin causes dose-dependent pain, allodynia, and hyperalgesia in humans, *Journal of Investigative Medicine 54*, 238–244.
37. Gazerani, P., Andersen, O. K., and Arendt-Nielsen, L. (2007) Site-specific, dose-dependent, and sex-related responses to the experimental pain model induced by intradermal injection of capsaicin to the foreheads and forearms of healthy humans, *Journal of Orofacial Pain 21*, 289–302.
38. Dougherty, P. M., Schwartz, A., and Lenz, F. A. (1999) Responses of primate spinomesencephalic tract cells to intradermal capsaicin, *Neuroscience 90*, 1377–1392.
39. Gilchrist, H. D., Allard, B. L., and Simone, D. A. (1996) Enhanced withdrawal responses to heat and mechanical stimuli following intraplantar injection of capsaicin in rats, *Pain 67*, 179–188.
40. Alvarez, P., Saavedra, G., Hernandez, A., Paeile, C., and Pelissier, T. (2003) Synergistic antinociceptive effects of ketamine and morphine in the orofacial capsaicin test in the rat, *Anesthesiology 99*, 969–975.
41. Pelissier, T., Pajot, J., and Dallel, R. (2002) The orofacial capsaicin test in rats: effects of different capsaicin concentrations and morphine, *Pain 96*, 81–87.
42. Chen, H.-S., He, X., Wang, Y., Wen, W.-W., You, H.-J., and Arendt-Nielsen, L. (2007) Roles of capsaicin-sensitive primary afferents in differential rat models of inflammatory pain: a systematic comparative study in conscious rats, *Experimental Neurology 204*, 244–251.
43. De Castro Costa, M., De Sutter, P., Gybels, J., and Van Hees, J. (1981) Adjuvant-induced arthritis in rats: a possible animal model of chronic pain, *Pain 10*, 173–185.
44. Millan, M. J., Czlonkowski, A., Pilcher, C. W., Almeida, O. F., Millan, M. H., Colpaert, F. C., and Herz, A. (1987) A model of chronic pain in the rat: functional correlates of alterations in the activity of opioid systems, *Journal of Neuroscience 7*, 77–87.
45. Coderre, T. J. and Wall, P. D. (1987) Ankle joint urate arthritis (AJUA) in rats: an alternative animal model of arthritis to that produced by Freund's adjuvant, *Pain 28*, 379–393.
46. Baek, Y. H., Choi, D. Y., Yang, H. I., and Park, D. S. (2005) Analgesic effect of electroacupuncture on inflammatory pain in the rat model of collagen-induced arthritis: mediation by cholinergic and serotonergic receptors, *Brain Research 1057*, 181–185.
47. Wu, J., Lin, Q., Lu, Y., Willis, W. D., and Westlund, K. N. (1998) Changes in nitric oxide synthase isoforms in the spinal cord of rat following induction of chronic arthritis, *Experimental Brain Research 118*, 457–465.
48. Barton, N. J., Strickland, I. T., Bond, S. M., Brash, H. M., Bate, S. T., Wilson, A. W., Chessell, I. P., Reeve, A. J., and McQueen, D. S. (2007) Pressure application measurement (PAM): a novel behavioural technique for measuring hypersensitivity in a rat model of joint pain, *Journal of Neuroscience Methods 163*, 67–75.
49. Butler, S. H., Godefroy, F., Besson, J. M., and Weil-Fugazza, J. (1992) A limited arthritic model for chronic pain studies in the rat, *Pain 48*, 73–81.
50. Sluka, K. A., Bailey, K., Bogush, J., Olson, R., and Ricketts, A. (1998) Treatment with either high or low frequency TENS reduces the secondary hyperalgesia observed after injection of kaolin and carrageenan into the knee joint, *Pain 77*, 97–102.
51. Sluka, K. A. and Westlund, K. N. (1993) Behavioral and immunohistochemical changes in an experimental arthritis model in rats, *Pain 55*, 367–377.
52. Ji, G., Fu, Y., Ruppert, K. A., and Neugebauer, V. (2007) Pain-related anxiety-like behavior requires CRF1 receptors in the amygdala, *Molecular Pain 3*, 13.
53. Schaible, H. G. and Schmidt, R. F. (1985) Effects of an experimental arthritis on the sensory properties of fine articular afferent units, *Journal of Neurophysiology 54*, 1109–1122.
54. Dougherty, P. M., Sluka, K. A., Sorkin, L. S., Westlund, K. N., and Willis, W. D. (1992) Neural changes in acute arthritis in monkeys. I. Parallel enhancement of responses of spinothalamic tract neurons to mechanical stimulation

and excitatory amino acids, *Brain Research. Brain Research Reviews 17*, 1–13.

55. Radhakrishnan, R., Moore, S. A., and Sluka, K. A. (2003) Unilateral carrageenan injection into muscle or joint induces chronic bilateral hyperalgesia in rats, *Pain 104*, 567–577.
56. Martins, M. A., De Castro Bastos, L., and Tonussi, C. R. (2006) Formalin injection into knee joints of rats: pharmacologic characterization of a deep somatic nociceptive model, *Journal of Pain 7*, 100–107.
57. Roveroni, R. C., Parada, C. A., Cecília, M., Veiga, F. A., and Tambeli, C. H. (2001) Development of a behavioral model of TMJ pain in rats: the TMJ formalin test, *Pain 94*, 185–191.
58. Okuda, K., Nakahama, H., Miyakawa, H., and Shima, K. (1984) Arthritis induced in cat by sodium urate: a possible animal model for tonic pain, *Pain 18*, 287–297.
59. Sluka, K. A. and Wright, A. (2001) Knee joint mobilization reduces secondary mechanical hyperalgesia induced by capsaicin injection into the ankle joint, *European Journal of Pain 5*, 81–87.
60. Lam, F. Y. and Ferrell, W. R. (1991) Neurogenic component of different models of acute inflammation in the rat knee joint, *Annals of the Rheumatic Diseases 50*, 747–751.
61. Reeh, P. W. and Steen, K. H. (1996) Tissue acidosis in nociception and pain, *Progress in Brain Research 113*, 143–151.
62. Steen, K. H., Reeh, P. W., Anton, F., and Handwerker, H. O. (1992) Protons selectively induce lasting excitation and sensitization to mechanical stimulation of nociceptors in rat skin, in vitro, *Journal of Neuroscience 12*, 86–95.
63. Sluka, K. A., Kalra, A., and Moore, S. A. (2001) Unilateral intramuscular injections of acidic saline produce a bilateral, long-lasting hyperalgesia, *Muscle & Nerve 24*, 37–46.
64. Ledeboer, A., Mahoney, J. H., Milligan, E. D., Martin, D., Maier, S. F., and Watkins, L. R. (2006) Spinal cord glia and interleukin-1 do not appear to mediate persistent allodynia induced by intramuscular acidic saline in rats, *Journal of Pain 7*, 757–767.
65. Diehl, B., Hoheisel, U., and Mense, S. (1988) Histological and neurophysiological changes induced by carrageenan in skeletal muscle of cat and rat, *Agents and Actions 25*, 210–213.
66. Kehl, L. J., Trempe, T. M., and Hargreaves, K. M. (2000) A new animal model for assessing mechanisms and management of muscle hyperalgesia, *Pain 85*, 333–343.
67. Beyreuther, B. K., Geis, C., Stöhr, T., and Sommer, C. (2007) Antihyperalgesic efficacy of lacosamide in a rat model for muscle pain induced by TNF, *Neuropharmacology 52*, 1312–1317.
68. Dina, O. A., Green, P. G., and Levine, J. D. (2008) Role of interleukin-6 in chronic muscle hyperalgesic priming, *Neuroscience 152*, 521–525.
69. Ro, J. Y., Capra, N., and Masri, R. (2003) Development of a behavioral assessment of craniofacial muscle pain in lightly anesthetized rats, *Pain 104*, 179–185.
70. Ro, J. Y. (2005) Bite force measurement in awake rats: a behavioral model for persistent orofacial muscle pain and hyperalgesia, *Journal of Orofacial Pain 19*, 159–167.
71. Nozaki-Taguchi, N. and Yaksh, T. L. (1998) A novel model of primary and secondary hyperalgesia after mild thermal injury in the rat, *Neuroscience Letters 254*, 25–28.
72. Wang, S., Lim, G., Yang, L., Zeng, Q., Sung, B., Jeevendra Martyn, J. A., and Mao, J. (2005) A rat model of unilateral hindpaw burn injury: Slowly developing rightwards shift of the morphine dose-response curve, *Pain 116*, 87–95.
73. Hylden, J. L., Thomas, D. A., Iadarola, M. J., Nahin, R. L., and Dubner, R. (1991) Spinal opioid analgesic effects are enhanced in a model of unilateral inflammation/hyperalgesia: possible involvement of noradrenergic mechanisms, *European Journal of Pharmacology 194*, 135–143.
74. Pedersen, J. L. and Kehlet, H. (1998) Secondary hyperalgesia to heat stimuli after burn injury in man, *Pain 76*, 377–384.
75. Chassaing, C., Schmidt, J., Eschalier, A., Cardot, J. M., and Dubray, C. (2006) Hyperalgesia induced by cutaneous freeze injury for testing analgesics in healthy volunteers, *British Journal of Clinical Pharmacology 61*, 389–397.
76. Bishop, T., Hewson, D. W., Yip, P. K., Fahey, M. S., Dawbarn, D., Young, A. R., and McMahon, S. B. (2007) Characterisation of ultraviolet-B-induced inflammation as a model of hyperalgesia in the rat, *Pain 131*, 70–82.
77. Harrison, G. I., Young, A. R., and McMahon, S. B. (2004) Ultraviolet radiation-induced inflammation as a model for cutaneous hyperalgesia, *Journal of Investigative Dermatology 122*, 183–189.
78. Chacur, M., Longo, I., Picolo, G., Gutiérrez, J. M., Lomonte, B., Guerra, J. L., Teixeira, C. F. P., and Cury, Y. (2003) Hyperalgesia induced by

Asp49 and Lys49 phospholipases A2 from Bothrops asper snake venom: pharmacological mediation and molecular determinants, *Toxicon 41*, 667–678.

79. Maier, S. F., Wiertelak, E. P., Martin, D., and Watkins, L. R. (1993) Interleukin-1 mediates the behavioral hyperalgesia produced by lithium chloride and endotoxin, *Brain Research 623*, 321–324.

80. Watkins, L. R., Wiertelak, E. P., Goehler, L. E., Mooney-Heiberger, K., Martinez, J., Furness, L., Smith, K. P., and Maier, S. F. (1994) Neurocircuitry of illness-induced hyperalgesia, *Brain Research 639*, 283–299.

Chapter 3

Animal Models of Visceral Pain

Karin N. Westlund

Abstract

The study of visceral pain is of high clinical relevance and the findings more directly translational in the search for analgesic agents. Early studies of nerve recordings after acute visceral nerve activation with (1) mechanical distension of hollow organs such as the colon or esophagus, (2) chemical irritation, and/or (3) inflammation have provided relevant information about normal, typically silent visceral afferents and evoked behavioral responses. Clinically relevant information about visceral pain has been reported in studies utilizing intact animal models of inflammatory, diabetic, neuropathic, cancer, chemotherapy-induced and other injury-related visceral pain conditions. More recently, animal models designed to study mechanisms signaling the transitional stages from acute to chronic visceral pain are providing information relevant to the development of drug therapies for reducing visceral pain in patients with these conditions. As an example, a visceral pain model of chronic pancreatitis is induced with an alcohol and high-fat diet in rat that persists for 2 months. Increased sensitization, inflammation, and pancreatic tissue disruption typical in the model are reversed by overexpression of met-enkephalin by a herpes simplex viral (HSV) preproenkephalin vector. The therapeutic effects are most likely attributable to opioid receptors located both in the central and peripheral nervous systems, as well as on peripheral cells. Advantages of using (HSV) vectors for therapeutic gene delivery rather than other viral delivery vectors (lentivirus, adenovirus) in certain clinical settings to treat visceral pain are discussed.

1. Introduction

Visceral pain is of great clinical concern, constituting the majority of pain treated by the medical community. Unfortunately, considerably fewer studies of visceral pain have been done, and fewer animal models have been devised for its study. This is partly because the study of visceral pain is generally much more difficult than studying cutaneous pain. Nevertheless, the study of visceral pain is of high clinical relevance and the findings more directly translational (91). The cursory assumption that mechanisms of cutaneous pain, neurotransmitters, and pathways of neural transmission for visceral pain are similar to somatic pain is not true, particularly for structures innervated by lumbosacral afferents (for reviews, see Buffington (23)).

Chao Ma and Jun-Ming Zhang (eds.), *Animal Models of Pain*, Neuromethods, vol. 49,
DOI 10.1007/978-1-60761-880-5_3,

Visceral pain studies in animal models have provided important information about the similarities and significant differences from somatic pain, for example, in the neural pathways, transmitters, and mechanistic properties utilized. A most recent and highly relevant point of interest is the opiate responsivity differences between male and female rats that have been shown with a visceral pain colorectal distension model (84). Endogenous opioid levels, opiate binding, and pharmacological effects at the spinal and supraspinal levels are detected that account for the sex differences. These same differences may account for therapeutic efficacy differences for opiate therapy in male and female patients. The data imply that important pharmacological data is yet to be discovered using visceral pain models, particularly chronic visceral pain models.

2. Animal Models of Visceral Pain

Early studies of visceral afferent activation began as did study of somatic afferents by stimulating nerves innervating internal organs (viscera) electrically or chemically in isolated nerve preparations from normal animals. Distension of hollow organs such as the colon or esophagus has also provided relevant information about normal visceral afferent transmission and behavioral responses. More recently, clinically relevant information about visceral pain has been reported in studies with intact animal models of inflammatory, diabetic, neuropathic, cancer, chemotherapy agent, and other injury-related visceral pain. Animal models designed to study mechanisms signaling the transitional stages from the acute to the chronic visceral pain condition are also providing information relevant to the development of drug therapies for reducing visceral pain in patients with these conditions.

Unfortunately, most models of visceral pain existing in the literature focus primarily on acute visceral nerve activation using (1) mechanical distension, (2) chemical irritation, and/or (3) inflammation. Few animal models of cancer pain have been forthcoming (120). While stimulation or perineural invasion of nerves in vivo has been the primary preparation for study of visceral pain, excised tissue preparations have been utilized successfully for colon (111). Better models of chronic visceral pain are clearly needed in addition to the acute nerve studies.

2.1. Mechanical Distension

Classic models for study of visceral afferent nerves include overfilling the bladder with fluid and pressure distension of the colon, ureter, duodenum, or esophagus (27, 45, 89, 119, 145). Most afferent nerve fibers of the S1 dorsal root identified by electrical stimulation also respond to direct mechanical or chemical stimulation of the urinary bladder, colon, or the anal mucosa. With these methods, it has been determined that of the many pelvic afferent nerve fibers responding to urinary bladder distension pressure (80 mmHg) (38–44% of the nerve total), 61% of the responding afferents are unmyelinated C-fibers, and 39% are thinly myelinated A delta fibers (119).

Interruption of the nerve or the dorsal column nucleus eliminates noxious visceral input to the thalamus and pain-related behavior after BK stimulation of pancreas (52, 53), as it does with colon distension (3, 5).

2.2. Chemical Irritation

Pancreatic, bladder, colon, and other visceral afferent fibers are sensitive to algesic chemicals such as turpentine (25%), mustard oil (2.5%), capsaicin (1%), and bradykinin (BK), allowing acute behavioral studies as well as recording of visceral nerves and central neuronal activation (63, 70, 76, 138). Behavioral studies indicate that mustard oil produces sensitized responses that begin immediately, persist, peak by 3 days. It produces longer-term, IBS-like, accelerated upper GI transit rates in mice (61).

BK is produced naturally during inflammation and excitation of nociceptors (63) and BK levels in the blood are known to increase quickly after induction of pancreatitis (150). Bradykinin is released both by damaged tissue and by infiltrating inflammatory cells. Lim et al. (69) report that BK produces pain similar to clinical visceral pain in 80% of human subjects. Direct application of BK has been shown to be an effective noxious stimulus in the pancreas activating responses in autonomic, dorsal column, or thalamic neurons (53, 94, 95, 136). In previous studies, assessment of the physiological responses of splanchnic afferent nerve fibers and spinal cord cells that respond to pancreatic afferent stimulation with BK in animals with pancreatitis have been compared to responses of normal animals (unpublished). In the published studies, application of BK onto the surface of the pancreas produced extracellular responses from cells of the dorsal column nucleus and ventral posterolateral nucleus of the thalamus, accompanied by mild autonomic responses. Pain-related behavioral responses are also observed after BK infusion directly into the pancreatic duct to simulate the pain of pancreatic ductal stenosis (53). Interruption of the pathway or the dorsal column nucleus significantly reduces pain-related behaviors and noxious visceral input to the thalamus after BK stimulation of pancreas (52, 53).

2.2.1. Animal Models of Bladder Pain

Animal models of cystitis are numerous. Instillation of noxious chemicals into the bladder produces increases in spinal Fos and a transient nerve activation that has been recorded in many studies. Early studies utilized instillation of mustard oil or turpentine for acute activation of urinary bladder afferents (44). A persistent bladder irritation model is produced by cyclophosphamide (50–74 mg/kg given i.p. for 2 weeks or on Day 1, Day 4) (18, 67, 149). Even more chronic irritation of the bladder in mice was produced in studies by the Sabans (114). Their studies indicated that the bladder responses to inflammation elicit upregulation of pain-related receptors on nociceptors that play a role in chronic cystitis pain. The chronic irritation of the bladder by algogens such as substance P, LPS, or an antigen after weekly injections of agents that increase IgE, provide gene expression profiles with distinct differences between acute and chronic bladder inflammation. Their studies find that NK1 receptors, mast cells, and bacterial products all play a critical role in allergic cystitis. These studies are a strong reminder that study of acute preparations may be used to characterize normal nerve function, but chronic visceral pain animal models are required to understand clinically relevant pathophysiology.

2.2.2. Animal Models of Pancreatic Pain

There is very little information available concerning mechanisms of nociceptive transmission from pancreas. However, several animal models of acute pancreatitis pain have been characterized in rats. A model of acute pancreatitis taken from Merkord and Hennighausen (80) and Merkord et al. (81) involves a tail vein injection of dibutyltin dichloride (DBTC, 8 mg/kg), an active compound of some paint thinners and plastics which is also used as a biocide in agriculture.

Pancreatic damage in the DBTC pancreatitis model peaks at 7 days and is maintained through 21 days (133). Serum parameters that serve as markers of pancreatitis remain elevated during the first week. With this model, it was determined that mechanical and thermal allodynia are present along the abdomen and persist through the first week. Characterization of other models of clinically relevant pancreatitis have been developed by pancreatic injections of trinitrobenzene sulfonic acid (TNBS) (99) or capsaicin (50), or even i.p. injection of L-arginine (48). These models have been used to determine that activation of afferent fibers containing substance P and calcitonin gene-related peptide (CGRP) and innervating the pancreas are TRPV1 and proteinase-activated receptor mediated (50, 141). The mice develop histological and immunological responses resembling severe acute bouts of human pancreatitis (12) which are aggravated by alcohol ingestion (100). These animal studies are relevant to the important suggestion that clinically used contrast media have a pH low enough to be contributing to further neurogenic inflammation in patients with pancreatitis (92).

While acute pancreatitis pain can be effectively treated clinically, chronic pancreatitis often produces a life-long debilitating pain state for many individuals. Animal models relevant to chronic pancreatitis and painful pancreatic cancer have been more difficult to produce. Rats appear to be particularly resistant to attempts to model this type of visceral pain. However, one model described is relatively simple if initiated under very specific conditions. This high-fat and alcohol-induced pancreatitis model persists through 8 weeks as published (148). The model can be utilized with Sprague Dawley or Fisher F344 rats. Provided here is data for

Hargreaves heat sensitivity (abdomen), von Frey mechanosensitivity, and hotplate tests indicating its usefulness for assessing a clinically relevant therapy over an 8-week time course in Fisher 344 rats (Fig. 1).

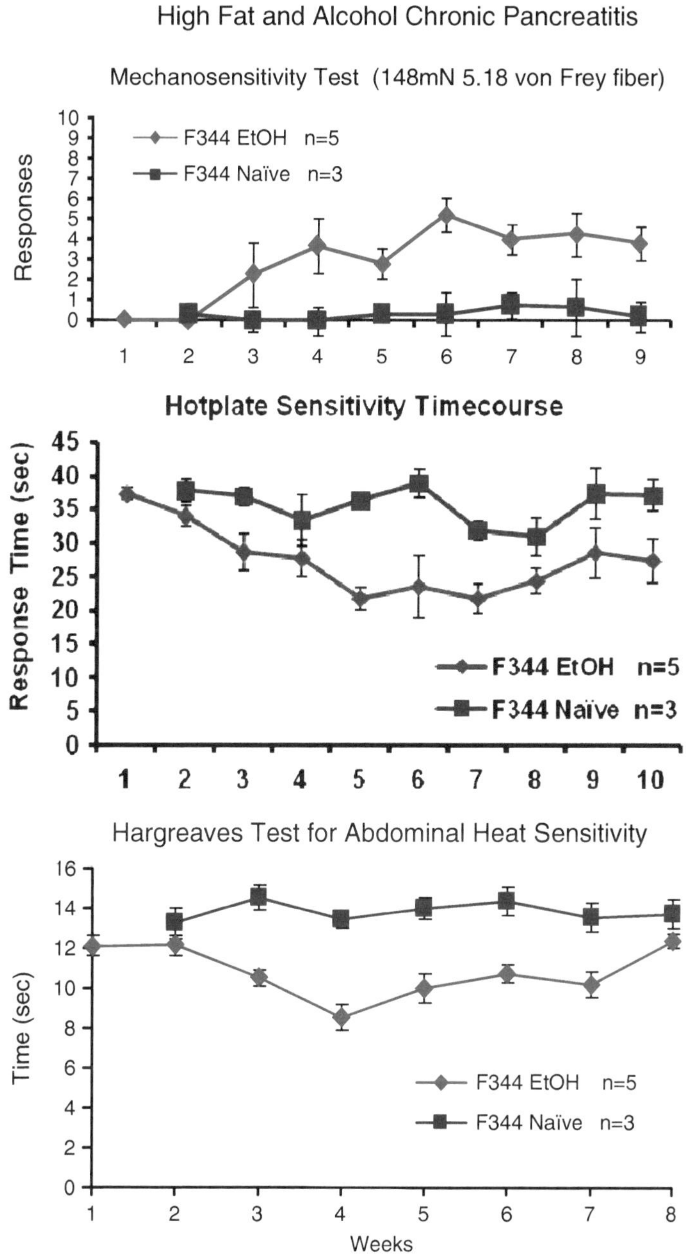

Fig. 1. Heat, mechanical, and hotplate hypersensitivity in a chronic pancreatitis model in rats. A chronic model of pancreatitis is produced in young Fisher 344 rats fed a high-fat and alcohol diet. The animals develop hypersensitivity to abdominal heat and von Frey fiber stimulation, as well as hotplate (50°C) sensitivity. The time course of the sensitivity is shown persisting over 6–7 weeks compared to responses of naïve animals fed normal chow.

Experimental studies have shown that pancreatic afferents reach the spinal cord and enter Lissauer's tract where they may terminate in the spinal segment in which they enter the spinal cord or within one or two segments rostrally or caudally (127). Some of the processes terminate on cells in the dorsal commissural nucleus (DCN) of lamina X that is involved in visceral processing. In a previous retrograde transport study from pancreas of rats, Sharkey et al. (121) showed retrogradely labeled cell bodies in the DRGs. The majority of these labeled cells were localized in the lower thoracic spinal ganglia.

3. Contributions of Animal Models to the Study of Visceral Pain

Key to understanding mechanisms responsible for visceral pain are the animal models for its study.

3.1. Animal Models Reveal Mechanisms of Visceral Afferent Activation

Studies in normal animals have revealed that primary afferent nerve fibers innervating viscera are polymodal mechanoreceptors (A delta fibers) and nociceptors (C-fibers), that is, they can respond to mechanical, thermal, and chemical stimuli. Afferent functions have been studied in hollow organs such as the esophagus, gall bladder, stomach, urinary bladder, colon, and uterus, since these organs can be distended with fluids or by air-filled balloons. Responses of visceral afferents have been shown to undergo profound changes when injured, when stimulated electrically or chemically, or when inflamed. Chemical stimulation of mechanically insensitive visceral afferents has been studied after application of algogenic agents, such as capsaicin, bradykinin, prostaglandins, leukotrienes, serotonin, histamine, and free radicals.

Numerous studies have shown that processing of information about nocifensive stimuli is altered peripherally or centrally, but typically, when it occurs at both sites, it can then develop into hypersensitive states. Peripherally, alterations corresponding to a developing sensitization of primary visceral afferents occur especially in the C-fiber polymodal nociceptors. Under normal physiological conditions, there is little activity in most visceral afferents. In response to injury or inflammation, recruitment of nociceptors innervating internal structures that previously had remained "silent" with disuse is highly typical of many visceral afferents (22, 74). It is the excessive and continuing activation of normally silent visceral afferents that provides the pathological stimulus for sensitization and clinical internal organ pain. The "awakening" of silent afferents was first reported for knee joint afferents in cat (116), and silent afferent studies have been reproduced in visceral afferents. It is estimated that 90% of the unmyelinated (C-fiber) bladder afferents are silent (reviewed in 82). Response of bladder afferents to overfilling the bladder occurs in 10% of silent afferents within 5 min (44). Response to chemical irritants in this acute study occurred rapidly in 45% of bladder afferents and dissipated upon removal of the irritant, while the other afferents (55%) remained silent.

3.2. Visceral Afferent Activation of Spinal Cord Neurons

When visceral afferents are stimulated excessively and/or over long periods of time by maximal distension, chemicals, inflammation, and/or infection, the nociceptive visceral afferent discharge then evokes profound central changes. A vast array of plastic changes have been observed in spinal cord neurons, including altered neuronal responses (39, 40, 90, 93,129) and altered expression of the early intermediate c-fos gene product, Fos (134). Prolonged noxious stimulation of the viscera evokes increases in the excitability of viscerosomatic neurons in the spinal cord which also typically are silent (28, 110). Such changes are very selective and highly organized as they occur initially only in those viscerosomatic cells that are driven by the conditioning visceral stimuli. However, animal studies have also shown that neurogenic spread of sensitization occurs from one viscera to another, as well as from viscera to somatic structures, e.g., "referred pain."

3.3. Animal Studies of Visceral Afferent Activation of Central Neurons and "Referred Pain"

It is well known that somatic and visceral pain differ substantially in perception. While somatic pain provides discrete, well localized, and typically a "sharp" pain sensation, visceral pain is a diffuse, burning, and difficult-to-localize sensation. Explanations for this and other differential features such as distinctive pharmacology have been reviewed by Jagger et al. (55).

Another classic feature of visceral pain is its referral to another region of the body, e.g., appendicitis and heart pain (47), often following a dermatomal pattern (77). It has been shown very clearly in animal studies that viscerosomatic sensitivity produced by excessive visceral stimulation leads to somatic sensitization (16, 88) and that the converse is also true (25). Activation of visceral afferents by distension of colon is monitored by the extent of abdominal muscle contraction recorded with electromyography. Animal visceral pain models employed in these studies included cyclophosphamide irritation, bladder or colon distension, and combinations with hindpaw inflammation [complete Freund's adjuvant (CFA, 25 μl, intraplantar)]. A visceral pain model devised for studies in mice also revealed viscerosomatic converge (64). In this model, mustard oil (1%) or capsaicin (0.1%) is administered by inserting a fine cannula into the colon via the anus. Visceral pain-related behaviors evident as somatic referred sensitization (licking abdomen, stretching, contractions of abdomen, etc.) are counted. A significant number of human studies of somatic referred visceral pain have also been reported. In human studies, Erichsen (see Head (47)) stated that pain from the urinary bladder is sometimes referred to the soles of the feet. Ness and colleagues (87) have reported sensitization as far as the knees in rats after repeated distensions of the bladder.

Using animal models, at least two mechanisms have been proposed for referred pain in the early animal model literature. The axon reflex theory was proposed by Sinclair, Weddell, and Feindel in 1948 (122) which included the notion that peripherally the same axons innervate both somatic and visceral targets. Hence, when one branch becomes sensitized, other branches may become sensitized. Convergence of afferent input at a central site that becomes an irritable focus is another mechanism proposed for the phenomena (73, 112). It is now assumed that both mechanisms are actively involved in referred pain. Studies using animal models have confirmed both hypotheses for visceral pain. In rats, colonic irritation sensitizes urinary bladder afferents to noxious mechanical and chemical stimuli. Responses of the pelvic nerve to algesic agents, capsaicin, bradykinin, and substance P, or to bladder distension are significantly enhanced in a colon inflammation model induced with colon afferent chemical irritation with TNBS (132). Since the enhanced responses are blocked by denervation, neurogenic sensitization is confirmed. Localization of single dorsal root ganglia with fluorescent dyes received simultaneously from ulnar nerve and heart (78) or colon and bladder (29) confirm convergence of function in peripheral nerves. Other examples of spinal neuronal convergence are also provided in the literature (103).

3.4. Visceral Activation Models Produce Central Sensitization

Continued activation of visceral afferents would allow transition from acute to chronic visceral pain states with plastic changes in spinal cord and supraspinal neuronal responses and endogenous pain-modulating systems. Analysis of the plastic changes that take place in the central nervous system (CNS) in response to persistent noxious stimuli has generated a good deal of interest due to obvious clinical relevance. Central sensitization begins with altered afferent input and the release of neuroactive chemicals in the spinal dorsal horn. Some neuroactive chemicals released in response to visceral pain models vary from those released by somatic nerve activation (for review see de Groat (32)). These include vasoactive intestinal polypeptide (VIP), peptide histidine isoleucine amide (PHI), cholecystokinin (CCK), and opioid peptides. Visceral afferent activation also releases peptide neurotransmitters in common with somatic afferents including substance P (SP) and CGRP. For example, SP is known to be directly involved in visceral inflammation, and the bladder is heavily innervated by SP-containing nerves.

Since many spinal neurons have both visceral and somatic input, the increase in excitability of spinal neurons produced by these neuroactive chemicals can play a role in the pathogenesis of hyperalgesia and allodynia manifested in somatic structures despite a visceral etiology (76, 93). One of the major consequences of this process is the expansion of the receptive fields which can lead to a "memory" of the initiating peripheral insult (46, 113). This can last several hours under experimental conditions (151). While mechanisms involved in central hyperexcitability have been best characterized in models of cutaneous nociception, reports involving activation of visceral afferents from urinary bladder, colon, esophagus, and gallbladder emphasize that similar mechanisms are at work in visceral sensitization (26, 65, 75, 110).

It is known clinically that sensitization in one structure can affect sensation arising from another structure with anatomically related spinal control neurons. This was shown, for example, with a visceral pain model produced by pseudorabies virus injection into a tail muscle which then travels retrogradely along the nerve to infect cells in the spinal cord (57). In this case, the infection from neurons involved in neural control of somatic structures near the colon through central spread to the bladder control cells produces a neurogenic bladder inflammation secondarily, i.e., the bladder inflammation is dependent on infection of somatic neural structures rather than frank infection of the bladder.

Studies of colon and pancreas afferent activation have been examined by our research group, revealing sensitization of neurons along routes we describe for visceral nociceptive transmission. Visceral nociceptive responses activate neurons along pathways traditionally associated with transmission of pain, but also activate supraspinal regions through pathways not previously considered, since under normal conditions no input is received along this route from visceral structures. However, distension and, in particular, inflammation of colon or pancreas can activate the postsynaptic dorsal column (PSDC) cells in the spinal cord and the input transmitted via a midline dorsal column pathway (3–5, 52, 53).

The PSDC pathway involved in visceral nociceptive transmission, for example, can undergo a sensitization process during inflammation of visceral organs (5, 35, 49). Previous studies in our laboratory using an acute model of pancreatic nociceptive stimulation have provided evidence that the nociceptive signals from the pancreas travel in the recently explored dorsal column pathway, a visceral pain relay pathway, to activate the thalamus (4, 5, 53, 137).

3.5. Visceral Nociception Models Increase c-Fos Expression

The early intermediate gene product Fos is an especially useful activation marker in visceral nociceptive models. In studies of visceral pain where the visceral organ or its nerves are highly activated or activation is prolonged, plastic changes observed range from altered neuronal responses (40, 93, 90, 129) to altered expression of the early intermediate c-fos gene product, Fos, at the level of the spinal cord (134). For example, prolonged noxious stimulation of the viscera evokes increases in the excitability of viscerosomatic neurons in the spinal cord which typically are silent (28, 110). Such changes are very selective and highly organized as they occur initially only in those viscerosomatic cells that are driven by the conditioning visceral stimuli. However, studies have also shown that neurogenic spread of sensitization occurs from one viscera to another, as well as from viscera to somatic structures, e.g., "referred pain."

Particularly with strong nocifensive visceral stimuli or in models of longer duration, activation of neurons occurs at both spinal and supraspinal levels. Expression of Fos protein is reported in spinal interneurons and projection neurons in rat spinal cord in response to noxious visceral stimulation such as intraperitoneal acetic acid, bladder or colorectal distension or inflammation, pancreas inflammation, and other visceral pain models (34, 68, 71, 79, 96, 130, 131, 148). The distribution differs from that of Fos expression after somatic pain. Fos-labeled cells defining sites of pain processing in models of renal/ureteral obstruction and pancreatitis are reported in laminae I, lateral IV–V, and medial VII and X (10, 71), rather than more laterally. Another recent study found that after cyclophosphomide-induced bladder inflammation, bladder distension now produced a greater number of spinal dorsal horn neurons expressing Fos protein (134). They also noted that the distribution pattern of Fos-labeled neurons in the lumbosacral spinal cord is altered after distension when the bladder is inflamed. In addition to increased numbers of Fos-labeled cells in the dorsal horn, after bladder inflammation a larger population of the Fos-labeled cells (45%) are then located around the central canal in the DCN (lamina X).

In our laboratory, a study has shown that nociceptive stimulation of the pancreas induces c-fos expression primarily in segments T7-T10 of the spinal cord, indicating that nociceptive information from the pancreas is processed mainly in those spinal segments (Fig. 2). Expression of Fos protein has also been localized in brainstem and higher brain sites after visceral nociceptive insult in a variety of animal models (109). Fos localization is frequently reported in visceral pain models at supraspinal levels and is most likely reflective of the persisting nature of the sensitization that occurs with many visceral pain models. For example, Fos is identified in the ventral reticular formation and dorsolateral pons in models of bladder or pancreas inflammation and colorectal distension (68, 131, 148) (Fig. 2). Supraspinal regions including the periaqueductal gray, dorsal raphe, medial thalamus, and the central nucleus of the amygdala have also been shown to express Fos after cyclophosphamide-induced bladder pain (18, 67), after 2 months of pancreatitis or BK activation of pancreatic afferents (106, 148).

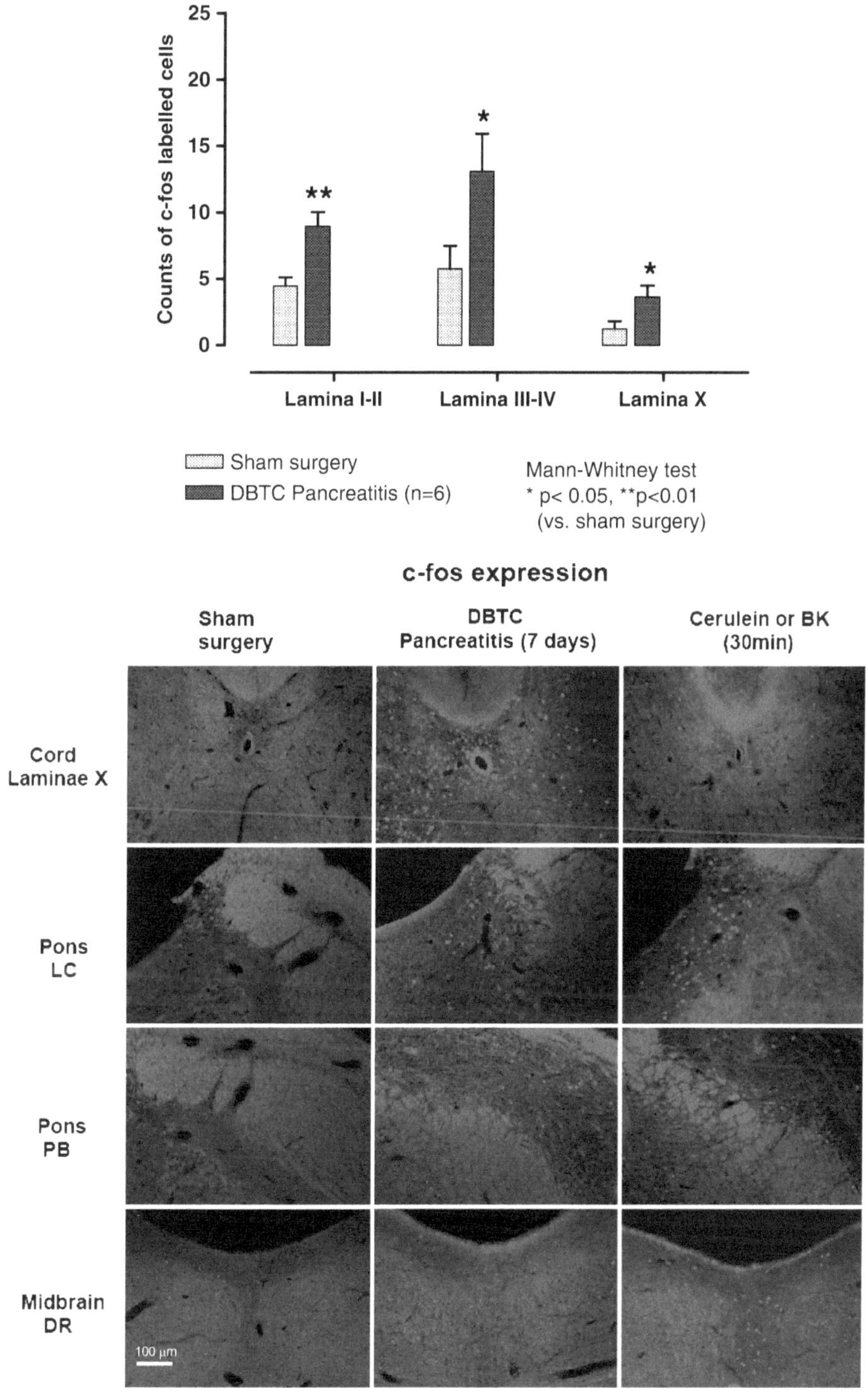

Fig. 2. c-Fos expression in chronic and acute pancreatitis. Activation of spinal cord and brainstem regions is indicated by c-Fos expression in various pancreatitis models, including after induction with DBTC (1 week) or after activation of pancreatic afferent fibers with cerulean or bradykinin. Cells with Fos-labeled nuclei are found in the lower thoracic cord superficial laminae I, II, in Lamina I-V, and in the laminae X region around the central canal shown. c-Fos in brainstem regions, including the locus coeruleus in the pons and the dorsal raphe in the midbrain are shown.

Thus, expression of Fos protein has proved to be a useful marker in various animal models of visceral insult for identification of CNS neurons responsive to enhanced nociception interpreted as visceral pain, while relatively little Fos protein is found in the spinal cord and other sites under normal conditions. Use of a polyclonal Fos antibody typically also allows visualization of Fos protein present after persisting visceral pain (71).

4. Pathways Transmitting Information About Noxious Visceral Stimulation Revealed in Animal Models

4.1. Visceral Afferents

Central terminals of visceral afferent nerves have been shown in guinea pigs to diffusely innervate spinal laminae I, II, IV, V, and X, whereas cutaneous afferents densely innervate only lamina I and II (127).

4.2. Postsynaptic Dorsal Column Pathway

There is increasing clinical and experimental evidence that the DC pathway, arising from lamina III and X PSDC cells, plays a principal role in central transmission of visceral nociceptive input. In clinical settings, patients with neurosurgical lesions of the midline dorsal column pathway have been relieved from the pain caused by pelvic cancer (13, 41, 49, 85, 86). After neurosurgery, the pain was completely relieved; large doses of morphine were no longer needed and the patient's vibratory, light touch, and proprioceptive sensation remained unchanged (85). The lesion site in patients was found, at postmortem examination, to interrupt only the most medial axons of the dorsal column (49). Recently, another study has been published in which it is reported that a number of patients were relieved from stomach cancer pain with a high thoracic dorsal column myelotomy situated at the lateral edge of the fasciculus gracilis (60). In experimental studies, Amassian (6–8) first reported that DC lesions at the cervical spinal levels in rabbits, cats, and dogs permanently abolished cortical activity after splanchnic nerve stimulation.

Electrophysiological studies have shown activity recorded from several sites along the DC-medial lemniscus pathway in response to stimulation of visceral afferents (1, 7, 14, 107, 108). Studies in our group provided evidence that DC pathways mediate nociceptive signals arising from the colon (3–5), the duodenum (35), and the pancreas (52, 53).

Anatomical studies have shown retrograde tracer injection into dorsal column nucleus and dorsal column midline itself labels a substantial number of PSDC cells in lamina III, IV, and X, particularly at thoracic and sacral spinal cord levels (49, 96). Many of the cells identified as PSDC cells after activation with noxious ureter distension were double labeled for Fos. Identified double-labeled spinothalamic tract cells were also localized in the same laminae, as well as in lamina I after noxious ureter distension.

Anterograde studies in our laboratory have shown that the PSDC cells send their uncrossed axons rostrally in the DC midline from sacral levels (49), while the thoracic level fibers travel viscerotopically between the gracile and cuneate fasciculi (49, 137). The dorsal column nuclei are primary termination sites of the PSDC pathway relaying the visceral input from activated PSDC cells in spinal cord. The third-order dorsal column nuclei neurons then relay the information to the thalamus contralaterally by way of the medial lemniscal pathway. As in patients with dorsal column lesions for pelvic cancer pain, dorsal column lesions interrupting axons ascending in the dorsal column midline block transmission of visceral nociceptive input in animal models of noxious pancreatic stimulation (52, 53), duodenal distension (35), and colorectal distension (3, 4) in rats and monkeys (2).

4.3. Spinoreticular and Spinothalamic Tract

Other major brain sites receive direct spinal input from lamina X, the spinal cord region around the central canal where neurons are responsive to noxious visceral input. This ascending pathway is revealed with tract tracing studies arising from cells located around the central canal in lamina VII and X. Bilateral input to supraspinal brain regions from lamina X travels in the ventral white matter near the ventral midline as an extension of the ventrolateral spinothalamic tract (137). Thus, while the ascending pathway is closely associated anatomically with the spinothalamic tract, it is far more bilateral than ascending spinothalamic tract input arising from lamina I, IV, and V. Another striking difference is that, like the lamina I thalamic projection, the lamina X axonal projections do not innervate the ventral posterolateral thalamus. Rather the lamina X projection innervates the intralaminar and medial thalamic nuclei, regions known to be responsive to both somatic and visceral input. Other supraspinal regions innervated by the ascending lamina X spinoreticular axons in our study (137) include the medullary reticular formation, raphe, periaqueductal gray, parabrachial nucleus, hypothalamus, central nucleus of the amygdala, and medial regions of the thalamus. These brain regions have been shown to be concerned with control of visceral and endocrine function, autonomic control, and affective perception of pain in a variety of studies using visceral pain models, including with fMRI 7 days after induction of DBTC pancreatitis (15, 56, 59, 137, 140). Noxious stimulation of the pancreas has been used to activate and reveal brain stem centers involved in pancreatic nociceptive processing. The activation marker, Fos, found in cells of these same brainstem regions, as well as the spinal cord, in a chronic alcohol-induced pancreatitis pain model is evidence of central sensitization in this model (148). The central nucleus of the amygdala has been shown to express c-fos after cyclophosphamide-induced bladder pain (18). Our gene therapy approach detailed below, which increases met-enkephalin in the spinal cord in a pancreatitis model, dramatically decreases Fos staining in supraspinal regions in animals with pancreatitis that have received the therapy (148).

5. Experimental Treatment Strategies to Reduce Visceral Pain Pancreatitis and Pancreatic Cancer Pain

Abdominal pain is the major presenting complaint of nearly all patients seeking medical attention with acute or chronic pancreatitis and those discovered to have pancreatic cancer. The annual incidence of new cases of acute pancreatitis reported in the US is 20 per 100,000 acute and 8 per 100,000 chronic pancreatitis cases. Severe pain was reported in over 70% of the 32,000 cancer patients who were expected to die in 2004 from pancreatic cancer, the 4th leading cause of cancer death overall. Most of the present knowledge about pancreatic pain has come from clinical work. For example, the effectiveness of blocks or sectioning of the sympathetic innervation, but not the vagus, indicates that afferents traveling with the sympathetic nerves are involved in transmitting nociceptive information from the pancreas (19, 33, 42, 54, 104, 126). Pancreatitis pain is usually epigastric in nature and may radiate to the right and /or left upper quadrant and to the back (19). Mechanical and thermal thresholds in these areas are decreased (30, 62).

While in some cases the cause of pancreatitis is unknown, the majority of cases are induced after prolonged alcohol consumption or exposure to industrial solvents/paint products. Pancreatitis is characterized by severe histopathological changes, such as the presence of inflammatory mediators, acinar atrophy, fat necrosis, intraductal hemorrhage, periductal fibrosis, and stromal proliferation (117). Elevated serum α-amylase and lipase levels serve as biochemical markers of acute pancreatitis (81, 123). Acute pancreatitis ranges from mild edematous conditions that usually heal without intervention to severe hemorrhagic necrotizing inflammation that is often fatal over a period of days as patients succumb to abdominal sepsis and multiorgan failure (118). The level of pain experienced by these patients is directly linked to decreased pancreatic functioning and increased length of stay during hospitalizations. In patient surveys, 32% of patients in chronic pain report being willing to try any new therapy for relief, and some may resort to suicide for this intractable pain state. Thus, the need to pursue novel pain relief strategies for pancreatitis and pancreatic cancer remains high.

5.1. Endothelins

Use of visceral pain models to develop pharmacological treatments for reducing visceral pain is an ultimate goal. For example, blockers of endothelin receptors A and B might be developed for treatment of pancreatic pain. The endothelins (ETs) are 21 amino acid peptides that are potent vasoconstrictors expressed by a variety of cell types including endothelial cells, macrophages, astrocytes, and neurons. In mammals, ETs produce their biological effects via activation of two receptor subtypes, the endothelin-A receptor (ET_AR) and the endothelin-B receptor (ET_BR) (98). It is known that the pancreas has abundant endothelin receptors and that levels are increased in inflammatory states (36). Abundant ET_AR after induction of DBTC pancreatitis is localized to neurons and ET_BR to satellite glia cells of the DRG (Fig. 3, Westlund, Vera-Portocarrero and Lu, unpublished). Endothelin receptor blockade affects the development of pancreatitis-associated microcirculatory failure, inflammation, and parenchymal injury (101, 102). Endothelin receptor blockade in severe acute pancreatitis leads to systemic enhancement of microcirculation, stabilization of capillary permeability, and improved survival rates (37). Localization on

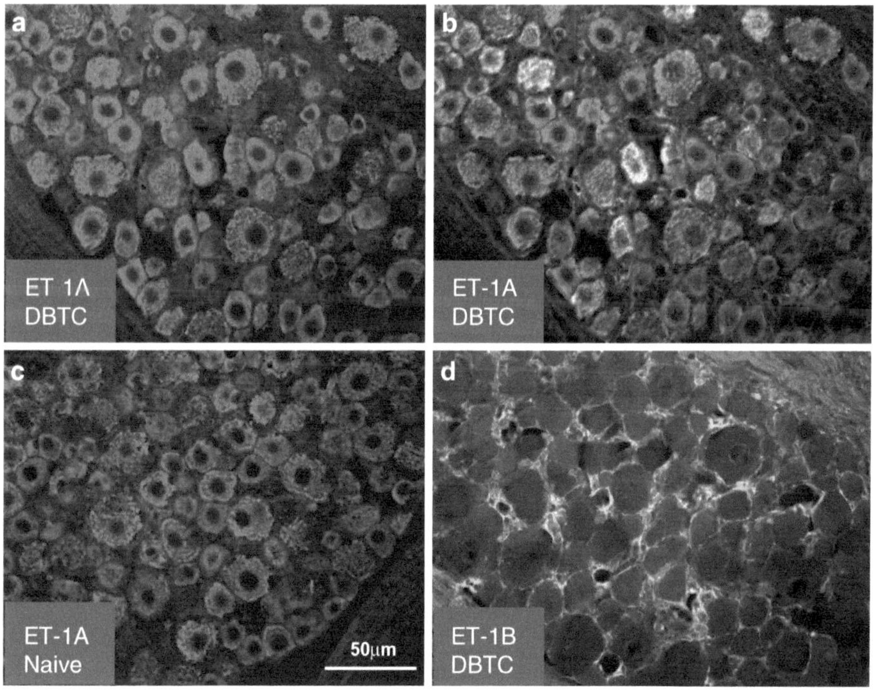

Fig. 3. Endothelin receptors in DRG of rats with DBTC-induced pancreatitis in comparison to normal rats. (**a**) Endothelin receptor A (ET-1A) in the thoracic DRG of rats with DBTC-induced pancreatitis. (**b**) ET-1A is costained with neuN (white) in DRG of multiple sizes. (**c**) ET-1A staining in thoracic DRG from naïve rats. (**d**) Endothelin receptor B (ET-1B) in the thoracic DRG of rats with DBTC-induced pancreatitis costaining with glial fibrillary acidic protein (GFAP, white) identifying the satellite cells.

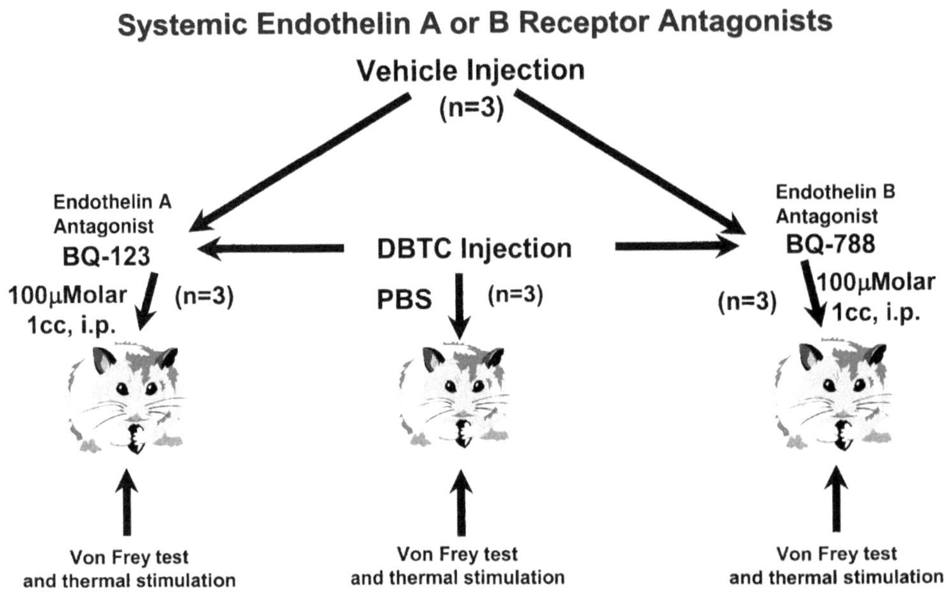

Fig. 4. Systemic treatment of animals with DBTC-induced pancreatitis with endothelin antagonists.

the peripheral sensory nerves and peripheral glia implies that ETs may be directly involved in signaling nociceptive events (98). We find that systemic treatment of animals with the DBTC pancreatitis (Fig. 4) with blockers of either ET_AR or ET_BR reduces mechanical, hotplate, and thermal sensitization responses tested on the abdominal skin (Fig. 5). The endothelin receptors (ETRs) have also been implicated in bone cancer pain (97). These findings suggest that agents reducing ETs may be effective in treatment of chronic visceral pain.

5.2. Opiates

Few studies to date report effective decreases in ongoing visceral pain other than with opiates. Activation of opiate receptors leads to potent analgesia (89,152), and opiates remain the primary therapeutic agent despite significant side effects and development of tolerance. A recent report provides evidence, however, of significantly reduced effectiveness of opiates in female rats during colorectal distension, reduced mu opioid receptor protein levels and radioligand binding within the pontine parabrachial nucleus of female rats in comparison to males. The data reported indicate that there are profound sex differences in how a noxious visceral stimulus and opiates engage the spino-parabrachial pathway (84). Other researchers report effectiveness of combining morphine and bupivicaine for suppression of post-operative pain after gall bladder removal when morphine is applied spinally (38), implying that suppression of visceral nociceptive responses may be improved with combined therapies in animal models.

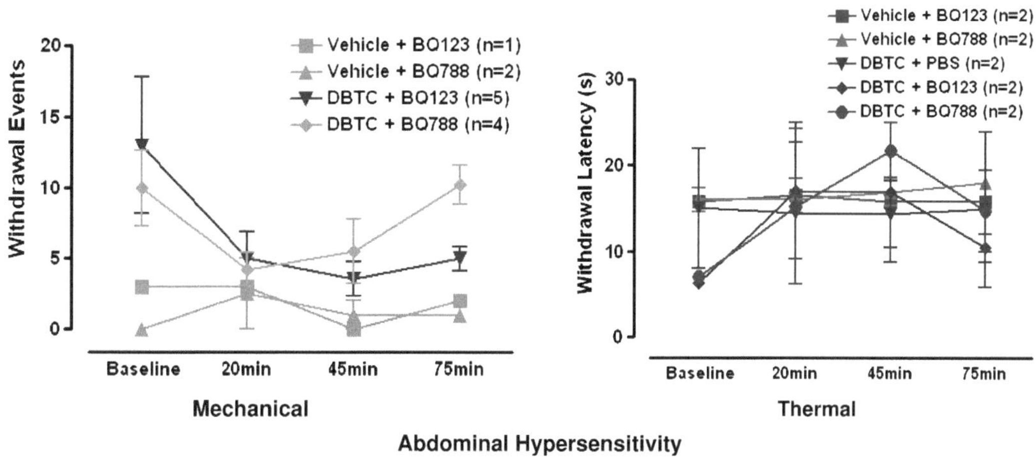

Fig. 5. Heat, mechanical, and hotplate hypersensitivity in DBTC-induced pancreatitis with and without endothelin antagonists. The ability of endothelin receptor antagonists to reverse the behavioral hypersensitivity present on Day 7 in rats with persistent DBTC pancreatitis was assessed. Endothelin receptor A (ET-1A) antagonist, (BQ-123, 1 cc, 300 μM, i.p.) significantly reversed the mechanical hypersensitivity assessed with von Frey filaments at 20 and 45 min after injection. The endothelin-B receptor antagonist (BQ-788) was only effective at 25 min. Both the endothelin-A (BQ123) and -B (BQ-788) receptor antagonists significantly reverse the thermal hypersensitivity assessed using the Hargreaves method at 20 and 45 min after injection. Neither agent affected vehicle injected animals.

Considerable improvements have been made in opioid medications, in term of bioavailability, half-life, transdermal delivery, breakthrough opioid combinations, and usage with enhancing drugs, such as tricyclic antidepressants. However, side effects and potential for opioid addiction remain a concern to the patient, health-care providers and the community. About 60% of patients with chronic pain have expressed fears regarding narcotic medication side effects and fear of addiction to current narcotic regimens. In a Partners Against Pain Survey of 1,000 patients in chronic pain, 50% of the patients reported difficulty >1 year in getting their pain under control and 78% of the patients reported that they would be willing to try new treatments.

5.3. Viral and Non-viral Vectors Generating Natural Opioids with Analgesic and Anti-inflammatory Potential

Gene therapy studies have successfully used herpes simplex viral 1 (HSV-1) transgene therapy to assess the antinociceptive effects of transduced opioids in numerous inflammatory models, including experimental models of acute and pancreatitis pain (21, 71, 72, 142, 146, 148). Gene therapy has several potential applications beyond replacement of normal genes or restitution of normal gene function. Gene therapy has also been proposed as a novel drug delivery system to express or possibly overexpress proteins in tissues. Two strategies have been reported for generating HSV-1 viral constructs for gene therapy. In one strategy, replication defective viral constructs have deletions of essential immediate early genes from the HSV-1 genome. In another strategy, replication conditional viruses are generated by insertion of the desired gene into the HSV-1 genes required for productive infections (i.e., thymidine kinase gene).

The HSV-1 viral vectors have an affinity for uptake by primary sensory neuronal endings (neurotropic), making them a preferable gene delivery construct for gene therapy analgesia. Wild-type herpes simplex virus type 1 (HSV-1) is a 154-kb neurotropic double-stranded DNA virus, containing 84 viral genes that constitute essential and nonessential genes based on in vitro studies. After natural primary cutaneous or mucosal inoculation, the virus undergoes lytic replication in infected epithelia. Viral particles are released at the site of the primary lesion, where they may enter sensory neurons whose axon terminals innervate the affected area. The HSV-1 nucleocapsid and tegument are carried naturally by retrograde axonal transport from the periphery to the DRG, where the virus may establish a life-long latent state (24, 125).

The recombinant viral vectors designed for gene therapy rendered replication conditional or defective do not produce a productive viral infection in vivo, but can persist for months despite negligible viral protein synthesis. These viral constructs establish a quiescent state similar to natural viral latency but cannot reactivate to cause an active infection in neuronal cells in vivo (24, 143, 145, 147).

In the numerous published studies demonstrating successful reduction of hypersensitivity in pain models without any reported side effects, application of the HSV-1 virus vector containing a human proenkephalin gene resulted in transmission by natural spread of incompetent viral protein through primary afferent fibers into the dorsal root ganglia. Replication conditional and replication defective viral constructs overexpressing met-enkephalin (71, 72, 144, 148) have been used to demonstrate effective reduction in nociceptive behaviors with no effect for control HSV-1-β -galactosidase-encoding vector or mock-infection model. Since we have evidence that the viral vectors are not active except in states of inflammation, we have used the vectors in both the short-term DBTC and the high-fat and alcohol-fed chronic pancreatitis models to examine potential for testing the efficacy of peptide production by the HSV-1 viral vector which would have obvious therapeutic advantages.

Subsequent protein expression from the proenkephalin or β-galactosidase gene (as a neutral control) can be visualized in the spinal cord (71, 72, 144, 148). Human proenkephalin-encoding HSV-1 viral vector reduced hyperalgesia by 60% in a polyarthritis model (21). The maximal response from the hindpaw application of HSV-1 enkephalin-coding viral vector was reported at 14 days (21). In addition, the experimental polyarthritis model provided radiographic evidence that the animals receiving the HSV enkephalin gene encoding viral vectors also sustained less joint destruction than the control animals after CFA injection. Both of the studies above and ours demonstrate that the opioid receptor antagonist, naloxone, delivered subcutaneously or intrathecally, can partially or completely restore nociception increases in animals infected with the HSV-1 enkephalin encoding vector (139).

We have found that the analgesia produced by HSV-Enk gene therapy also effectively reduces Fos staining occurring in spinal and supraspinal neurons in response to inflammation of the pancreas. Our data show that injection of the HSV-Enk viral construct directly into the pancreas effectively reduces or normalizes pain-related behaviors in rats in both acute (1 week) and chronic (7 weeks) experimental model of pancreatitis.

The mechanism of action for the reduction in pain-related behaviors may be related to enkephalin's influence on central and/or on peripheral opioid receptors. The opioid peptides are generated from three precursor genes, proenkephalin, prodynorphin, and POMC which have been conserved phylogenetically from invertebrate species (115). If the proenkephalin derived opioid peptides are released peripherally after HSV-1 administration, their effect would mimic the enhanced endogenous release of opiates from immune cells that invade the region of inflammation and modulate both pain and inflammatory parameters (124). While many of the opiate peptides are delivered by inflammatory cells drawn to the site of inflammation, some opioid peptides produced by local reactive cells may be more prominent in influencing nociceptive signaling. For example, proenkephalin gene product met-enkephalin is expressed in a 3:1 ratio relative to proenkephalin gene product, leu-enkephalin. The reason why some gene products from the opioid families have more efficient translational processing or are expressed in greater ratios is not completely understood (31). The proenkephalin viral construct was chosen for gene therapy studies since it is more prominent in influencing nociception. Clinical trials are currently underway with this viral vector (146).

Pharmacological block of HSV-Enk analgesia by administration of naloxone methiodide intrathecally to the spinal cord was possible since this isomer does not cross the blood–brain barrier allowing differentiation of the central versus peripheral effects of the gene product(s) produced by HSV-Enk viral vector infection. This finding suggests opioid mediation of the observed antinociceptive effects was at least, in part, due to the spinal release of opioids. Naloxone administration had no effect on hyperalgesia scores in animals infected with HSV-β-gal, nor did it affect the baseline responsiveness of the animals. These studies are compelling in that HSV-1 opioid gene delivery of the endogenous opioid peptide met-enkephalin potentially offers a significantly sustained response (up to 5 weeks) in the experimental models.

Even more surprising was our finding that the viral vector expression of met-enkephalin dramatically reduced the inflammatory component of the pancreatitis which no doubt contributed to the reduction in pain measures. Inflammatory processes were monitored after HSV-Enk generation of the opioid gene product and evidence found by histological examination for dramatic reduction of inflammatory and necrotic processes in the experimental group with excess overexpression of met-enkephalin. Since opioid receptors have been demonstrated not only in the CNS but also in the peripheral terminals of primary afferent neurons, it is likely that the HSV-Enk analgesic effect is mediated at central and peripheral sites. Activation of these receptors leads to potent analgesia (152). Experimental and clinical studies demonstrate potent analgesic effects after administration of low doses of opioid analgesics. Studies have reported that the efficacy and potency of peripheral opioid effects and other inhibitors are generally enhanced when drugs are administered during active inflammatory conditions (9, 20, 66, 124, 135, 152).

The decreased sensitization and inflammation effects are most likely induced by activation of both opioid receptors located in the central and peripheral nervous systems, as indicated by studies carried out with spinal and intrathecal administration of opioid agents. In the periphery, opioid receptors are expressed on a significant proportion of capsaicin-sensitive sensory fibers and sympathetic postganglionic terminals, where they may participate in modulation of nociceptive information under certain pathological conditions (152). The local administration of mu- and delta-opioid receptor agonists, at doses that show no systemic effect, have been shown to decrease plasma extravasation during peripheral inflammation (24–36%) (11, 17, 43, 51, 58, 128). It is likely that a similar mechanism reduces inflammatory signs in HSV-Enk-treated animals releasing enkephalin directly to pancreas.

Spinal opioid receptors are coupled to G-proteins that inhibit adenylate cyclase, inhibit calcium channels, and stimulate potassium channels, resulting in lower neuronal responsiveness (144). The opioid-mediated antihyperalgesia of HSV-Enk-infected animals mimics the effects of endogenous or intrathecally administered enkephalin. The lack of response of baseline hindpaw withdrawal responses in a cutaneous inflammation model suggests that the opioids affecting the antihyperalgesia may not be tonically released in the infected animals, but may be released only when there is a substantial activation of the afferents. These studies also provide additional evidence that hyperalgesia can be blocked without altering baseline nociception as reported previously for treatment in models of somatic pain (144). The studies we and others report establish the validity of use of HSV-based gene therapy in pain models, including pain of visceral origin.

There are several inherent advantages to using HSV-based viral constructs for foreign gene delivery in certain clinical settings, rather than other viral delivery models under study. Lentiviruses as viral vectors offer a very efficient infection and gene expression in activated cells and genome integration into host DNA for gene replacement therapy. They are the causal agent of cancers, autoimmune diseases, and acquired immune deficiency syndrome (AIDS). However, current lentiviral constructs result in systemic infection and cannot be contained. Likewise, the site of insertion of the lentivirus into the host genome cannot be controlled. Therefore, potential concerns exist regarding development of helper phenomena for unwanted expressions and genetic recombination with other lentiviruses, including those already integrated into the host genome.

Designing viral constructs from adenoviruses has also yielded promising results, but adenoviruses may not be advantageous since they can also have CNS involvement. Adenoviruses can also produce productive infections that can cause a host inflammatory response. Adenoviruses have a natural tropism for bronchial epithelium, liver, and pancreas. Adenoviruses can induce host inflammation above that already generated in target tissues and increase potential for extended injury and host innate immune response to the virus making repeated doses problematic (83).

HSV-1 viral constructs offer unique advantages for treating peripheral inflammation, as they naturally infect primary sensory neurons but they do not integrate into the host genome. These extra-chromosomal episomes can serve as an independent "minipump" source for protein synthesis in the neuronal cytoplasm. This is an advantage when delivery of supplemental amounts of protein might be therapeutic, as in pancreatic pain. Plus, as a DNA-based viral construct, (1) the rate of mutation and recombination in these HSV-1 constructs will be minimal to nonexistent since (2) these constructs do not enter into productive infection but enter latency. Further, the preliminary data suggest that HSV replication-deficient viral constructs do not generate proteins that would induce an amnestic response from the host, activating latent prior HSV-1 infections. These are very important properties of HSV-1 based viral vectors, as 70–90% of the human adult population has evidence of prior HSV-1 infection. This limited potential to generate a host response also improves the potential for using the HSV-based viral constructs for repeated dosing.

Unique advantages offered by HSV-1 viral vector delivery of proenkephalin expression products by the peripheral nerves revealed in experimental studies include:

1. The potential for delivery of analgesic peptides to both peripheral and central sites, optimizing the effects.
2. More effective abrogation of nociceptive responses with normalization of pain-related behavior near baseline levels.
3. Potential for positive impact on reduction of inflammation as shown in pancreatitis and arthritis studies.

4. HSV-1 viral constructs to be used are replication deficient/defective and will not incorporate into the host genome or become lytic even when injected directly into the CNS.
5. Focused delivery to the target organ allows a much lower viral titer (5–10× lower viral titer than in skin).
6. Potential for reduction/elimination of the use of narcotic drugs.

Thus, gene therapy with HSV-1 viral vectors may provide a novel palliative strategy for alleviating the unremitting pain present in pancreatic cancer, chronic pancreatitis, and other chronic visceral pain conditions.

Novel nonviral vector strategies are also on the horizon.

Acknowledgments

The work described in this chapter was carried out in collaboration with Drs. Ying Lu and Louis Vera-Portocarrero whose contributions are gratefully acknowledged. The technical contribution of Sarah Abshire to the chapter reference formatting and behavioral studies in the high-fat and alcohol pancreatitis study is also gratefully acknowledged. Research support was provided by NIH grant NS039041.

References

1. Aidar O., Geohegan W.A. and Ungewitter L.H. (1952) Splanchnic afferent pathways in the central nervous system. *J Neurophysiol* **15**, 131–8.
2. Al-Chaer E.D., Feng Y. and Willis W.D. (1999) Comparative study of viscerosomatic input onto postsynaptic dorsal column and spinothalamic tract neurons in the primate. *J Neurophysiol* **82**, 1876–82.
3. Al-Chaer E.D., Lawand N.B., Westlund K.N. and Willis W.D. (1996) Pelvic visceral input into the nucleus gracilis is largely mediated by the postsynaptic dorsal column pathway. *J Neurophysiol* **76**, 2675–90.
4. Al-Chaer E.D., Lawand N.B., Westlund K.N. and Willis W.D. (1996) Visceral nociceptive input into the ventral posterolateral nucleus of the thalamus: a new function for the dorsal column pathway. *J Neurophysiol* **76**, 2661–74.
5. Al-Chaer E.D., Westlund K.N. and Willis W.D. (1996) Potentiation of thalamic responses to colorectal distension by visceral inflammation. *NeuroReport* **7**, 1635–9.
6. Amassian V.E. (1951) Cortical representation of visceral afferents. *J Neurophysiol* **14**, 433–44.
7. Amassian V.E. (1951) Fiber groups and spinal pathways of cortically represented visceral afferents. *J Neurophysiol* **14**, 445–60.
8. Amassian V.E. (1952) Interaction in the somatovisceral projection system. *Res Publ Assoc Res Nerv Ment Dis* **30**, 371–402.
9. Antonijevic S., Mousa M., Schafer M. and Stein C. (1995) Perineurial defect and peripheral opioid analgesia in inflammation. *J Neurosci* **15**, 165–72.
10. Avelino A., Cruz F. and Coimbra A. (1997) Sites of renal pain processing in the rat spinal cord. A c-fos study using a percutaneous method to perform ureteral obstruction. *J Auton Nerv Syst* **67**, 60–6.
11. Barber A. (1993) Mu- and kappa-opioid receptor agonists produce peripheral inhibition of neurogenic plasma extravasation in rat skin. *Eur J Pharmacol* **236**, 113–20.
12. Barthet M., Dubucquoy L., Garcia S., Gasmi M., Desreumaux P., Colombel J.F., Grimaud J.C.,

Iovanna J. and Dagorn J.C. (2003) Pancreatic changes in TNBS-induced colitis in mice. *Gastroentérol Clin Biol* **27**, 895–900.

13. Becker R., Sure U. and Bertalanffy H. (1999) Punctate midline myelotomy. A new approach in the management of visceral pain. *Acta Neurochir (Wien)* **141**, 881–3.
14. Berkley K.J. and Hubscher C.H. (1995) Are there separate central nervous system pathways for touch and pain? *Nat Med* **1**, 766–73.
15. Bernard J.F., Huang G.F. and Besson J.M. (1994) The parabrachial area: electrophysiological evidence for an involvement in visceral nociceptive processes. *J Neurophysiol* **71**, 1646–60.
16. Bielefeldt K., Lamb K. and Gebhart G.F. (2006) Convergence of sensory pathways in the development of somatic and visceral hypersensitivity. *Am J Physiol Gastrointest Liver Physiol* **291**, G658–65.
17. Binder W., Machelska H., Mousa S., Schmitt T., Riviere P.J., Junien J.L., Stein C. and Schafer M. (2001) Analgesic and antiinflammatory effects of two novel kappa-opioid peptides. *Anesthesiology* **94**, 1034–44.
18. Bon K., Lanteri-Minet M., Michiels J.F. and Menetrey D. (1998) Cyclophosphamide cystitis as a model of visceral pain in rats: a c-fos and Krox-24 study at telencephalic levels, with a note on pituitary adenylate cyclase activating polypeptide (PACAP). *Exp Brain Res* **122**, 165–74.
19. Bonica J.J. (1990) Evolution and current status of pain programs. *J Pain Symptom Manage* **5**, 368–74.
20. Boyle D.L., Moore J., Yang L., Sorkin L.S. and Firestein G.S. (2002) Spinal adenosine receptor activation inhibits inflammation and joint destruction in rat adjuvant-induced arthritis. *Arthritis Rheum* **46**, 3076–82.
21. Braz J., Beaufour C., Coutaux A., Epstein A.L., Cesselin F., Hamon M. and Pohl M. (2001) Therapeutic efficacy in experimental polyarthritis of viral-driven enkephalin overproduction in sensory neurons. *J Neurosci* **21**, 7881–8.
22. Bueno L., Fioramonti J., Delvaux M. and Frexinos J. (1997) Mediators and pharmacology of visceral sensitivity: from basic to clinical investigations. *Gastroenterology* **112**, 1714–43.
23. Buffington C.A. (2001) Visceral pain in humans: lessons from animals. *Curr Pain Headache Rep* **5**, 44–51.
24. Burton E.A., Wechuck J.B., Wendell S.K., Goins W.F., Fink D.J. and Glorioso J.C. (2001) Multiple applications for replication-defective herpes simplex virus vectors. *Stem Cells* **19**, 358–77.
25. Cameron D.M., Brennan T.J. and Gebhart G.F. (2008) Hind paw incision in the rat produces long-lasting colon hypersensitivity. *J Pain* **9**, 246–53.
26. Cervero F. and Laird J.M. (1999) Visceral pain. *Lancet* **353**, 2145–8.
27. Cervero F. and Sann H. (1989) Mechanically evoked responses of afferent fibers innervation the guinea-pig's ureter: an in vitro study. *J Physiol* **412**, 245–266.
28. Cervero F., Laird J.M. and Pozo M.A. (1992) Selective changes of receptive field properties of spinal nociceptive neurones induced by noxious visceral stimulation in the cat. *Pain* **51**, 335–42.
29. Christianson J.A., Liang R., Ustinova E.E., Davis B.M., Fraser M.O. and Pezzone M.A. (2007) Convergence of bladder and colon sensory innervation occurs at the primary afferent level. *Pain* **128**, 235–43.
30. Cooperman A.M. (1981) Chronic pancreatitis. *Surg Clin North Am* **61**, 71–83.
31. Danielson P.B. and Dores R.M. (1999) Molecular evolution of the opioid/orphanin gene family. *Gen Comp Endocrinol* **113**, 169–86.
32. de Groat W.C. (1987) Neuropeptides in pelvic afferent pathways. *Experientia* **43**, 801–13.
33. Detakats G. and Walter L. (1947) The treatment of pancreatic pain by splanchnic nerve section. *Surg Gynecol Obstet*, **85**, 742–6.
34. Dinis P., Charrua A., Avelino A. and Cruz F. (2004) Intravesical resiniferatoxin decreases spinal c-fos expression and increases bladder volume to reflex micturition in rats with chronic inflamed urinary bladders. *BJU Int* **94**, 153–7.
35. Feng Y., Cui M., Al-Chaer E.D. and Willis W.D. (1998) Epigastric antinociception by cervical dorsal column lesions in rats. *Anesthesiology* **89**, 411–20.
36. Ferreira S.H., Romitelli M. and de Nucci G. (1989) Endothelin-1 participation in overt and inflammatory pain. *J Cardiovasc Pharmacol* **13**, Suppl 5, S220–2.
37. Foitzik T., Eibl G., Hotz H.G., Faulhaber J., Kirchengast M. and Buhr H.J. (2000) Endothelin receptor blockade in severe acute pancreatitis leads to systemic enhancement of microcirculation, stabilization of capillary permeability, and improved survival rates. *Surgery* **128**, 399–407.
38. Fujii Y., Toyooka H. and Tanaka H. (1998) Efficacy of thoracic epidural analgesia following laparoscopic cholecystectomy. *Eur J Anaesthesiol* **15**, 342–4.
39. Giamberardino M.A., Valente R., de Bigontina P. and Vecchiet L. (1995) Artificial ureteral

calculosis in rats: behavioural characterization of visceral pain episodes and their relationship with referred lumbar muscle hyperalgesia. *Pain* **61**, 459–69.

40. Giamberardino M.A., Valente R., Affaitati G. and Vecchiet L. (1997) Central neuronal changes in recurrent visceral pain. *Int J Clin Pharmacol Res* **17**, 63–6.
41. Gildenberg P.L. and Hirshberg R.M. (1984) Limited myelotomy for the treatment of intractable cancer pain. *J Neurol Neurosurg Psychiatry* **47**, 94–6.
42. Gorbitz C. and Leavens M.E. (1971) Alcohol block of the celiac plexus for control of upper abdominal pain caused by cancer and pancreatitis. Technical note. *J Neurosurg* **34**, 575–9.
43. Green P.G. and Levine J.D. (1992) Delta- and kappa-opioid agonists inhibit plasma extravasation induced by bradykinin in the knee joint of the rat. *Neuroscience* **49**, 129–33.
44. Häbler H.-J., Jänig W. and Koltzenburg M. (1990) Activation of unmyelinated afferent fibres by mechanical stimuli and inflammation of the urinary bladder in the cat. *J Physiol* **425**, 545–62.
45. Häbler H.J., Janig W. and Koltzenburg M. (1993) Myelinated primary afferents of the sacral spinal cord responding to slow filling and distension of the cat urinary bladder. *J Physiol* **463**, 449–60.
46. Haley J.E. and Wilcox G.L. (1992) Involvement of excitatory amino acids and peptides in the spinal mechanisms underlying hyperalgesia. In: *Hyperalgesia and Allodynia* (ed. by Willis WD, Jr.), pp. 281–94. Raven, New York.
47. Head H. (1893) On disturbances of sensation with special reference to the pain of visceral disease. *Brain* **16**, 1–133.
48. Hegyi P., Rakonczay Z., Jr., Sari R., Gog C., Lonovics J., Takacs T. and Czako L.L. (2004). Arginine-induced experimental pancreatitis. *World J Gastroenterol* **10**, 2003–9.
49. Hirshberg R.M., Al-Chaer E.D., Lawand N.B., Westlund K.N. and Willis W.D. (1996) Is there a pathway in the posterior funiculus that signals visceral pain? *Pain* **67**, 291–305.
50. Hoogerwerf W.A., Zou L., Shenoy M., Sun D., Micci M.A., Lee-Hellmich H., Xiao S.Y., Winston J.H. and Pasricha P.J. (2001) The proteinase-activated receptor 2 is involved in nociception. *J Neurosci* **21**, 9036–42.
51. Hong Y. and Abbott F.V. (1995) Peripheral opioid modulation of pain and inflammation in the formalin test. *Eur J Pharmacol* **277**, 21–8.
52. Houghton A.K., Kadura S. and Westlund K.N. (1997) Dorsal column lesions reverse the reduction of homecage activity in rats with pancreatitis. *NeuroReport* **8**, 3795–800.
53. Houghton A.K., Wang C.C. and Westlund K.N. (2001) Do nociceptive signals from the pancreas travel in the dorsal column? *Pain* **89**, 207–20.
54. Hurwitz A. and Gurwitz J. (1950) Relief of pain in chronic relapsing pancreatitis by unilateral sympathectomy. *Arch Surg* **61**, 379–86.
55. Jagger S.I., Scott H. and Rice A. (1999) Inflammation of the rat urinary bladder is associated with a referred thermal hyperalgesia which is nerve growth factor dependent. *Br J Anaesth* **83**, 442–9.
56. Jasmin L., Burkey A.R., Card J.P. and Basbaum A.I. (1997) Transneuronal labeling of a nociceptive pathway, the spino-(trigemino-)parabrachio-amygdaloid, in the rat. *J Neurosci* **17**, 3751–65.
57. Jasmin L., Janni G., Manz H.J. and Rabkin S.D. (1998) Activation of CNS circuits producing a neurogenic cystitis: evidence for centrally induced peripheral inflammation. *J Neurosci* **18**, 10016–29.
58. Joris J., Costello A., Dubner R. and Hargreaves K.M. (1990) Opiates suppress carrageenan-induced edema and hyperthermia at doses that inhibit hyperalgesia. *Pain* **43**, 95–103.
59. Katter J.T., Dado R.J., Kostarczyk E. and Giesler G.J., Jr. (1996) Spinothalamic and spinohypothalamic tract neurons in the sacral spinal cord of rats. II. Responses to cutaneous and visceral stimuli. *J Neurophysiol* **75**, 2606–28.
60. Kim Y.S. and Kwon S.J. (2000) High thoracic midline dorsal column myelotomy for severe visceral pain due to advanced stomach cancer. *Neurosurgery* **46**, 85–90.
61. Kimball E.S., Palmer J.M., D'Andrea M.R., Hornby P.J. and Wade P.R. (2005) Acute colitis induction by oil of mustard results in later development of an IBS-like accelerated upper GI transit in mice. *Am J Physiol Gastrointest Liver Physiol* **288**, G1266–73.
62. Kloppel G. (1990) Pathology of chronic pancreatitis and pancreatic pain. *Acta Chir Scand* **156**, 261–5.
63. Kress K. and Reeh P.W. (1996) Chemical excitation and sensitization in nociceptors. In: *Neurobiology of Nociceptors* (ed. by Belemonte C, Cervero F), pp. 258–97. Oxford University Press, New York.
64. Laird J.M., Martinez-Caro L., Garcia-Nicas E. and Cervero F. (2001) A new model of visceral pain and referred hyperalgesia in the mouse. *Pain* **92**, 335–42.
65. Laird J.M., Roza C. and Cervero F. (1997) Effects of artificial calculosis on rat ureter motility: peripheral contribution to the pain of ureteric colic. *Am J Physiol* **272**, R1409–16.
66. Lamigeon C., Bellier J.P., Sacchettoni S., Rujano M. and Jacquemont B. (2001)

Enhanced neuronal protection from oxidative stress by coculture with glutamic acid decarboxylase-expressing astrocytes. *J Neurochem* 77, 598–606.

67. Lanteri-Minet M., Bon K., de Pommery J., Michiels J.F. and Menetrey D. (1995) Cyclophosphamide cystitis as a model of visceral pain in rats: model elaboration and spinal structures involved as revealed by the expression of c-Fos and Krox-24 proteins. *Exp Brain Res* **105**, 220–32.
68. Lanteri-Minet M., Isnardon P., de Pommery J. and Menetrey D. (1993) Spinal and hindbrain structures involved in visceroception and visceronociception as revealed by the expression of Fos, Jun and Krox-24 proteins. *Neuroscience* **55**, 737–53.
69. Lim R.K., Miller D.G., Guzman F., Rodgers D.W., Rogers R.W., Wang S.K., Chao P.Y. and Shih T.Y. (1967) Pain and analgesia evaluated by the intraperitoneal bradykinin-evoked pain method in man. *Clin Pharmacol Ther* **8**, 521–42.
70. Longhurst J.C., Kaufman M.P., Ordway G.A. and Musch T.I. (1984) Effects of bradykinin and capsaicin on endings of afferent fibers from abdominal visceral organs. *Am J Physiol* **247**, R552–9.
71. Lu Y., McNearney T.A., Lin W., Wilson S.P., Yeomans D.C. and Westlund K.N. (2007) Treatment of inflamed pancreas with enkephalin encoding HSV-1 recombinant vector reduces inflammatory damage and behavioral sequelae. *Mol Ther* **15**, 1812–9.
72. Lu Y., McNearney T.A., Wilson S.P., Yeomans D.C. and Westlund K.N. (2008) Joint capsule treatment with enkephalin-encoding HSV-1 recombinant vector reduces inflammatory damage and behavioural sequelae in rat CFA monoarthritis. *Eur J Neurosci* **27**, 1153–65.
73. Mackenzie K.A. (1909) XXIII. Resection of the sciatic nerve: neuroplasty: end results. *Ann Surg* **50**, 295–312.
74. Mayer E.A. and Gebhart G.F. (1994) Basic and clinical aspects of visceral hyperalgesia. *Gastroenterology* **107**, 271–93.
75. McMahon S.B. (1997) Are there fundamental differences in the peripheral mechanisms of visceral and somatic pain? *Behav Brain Sci* **20**, 381–91; discussion 435–513.
76. McMahon S.B. and Abel C. (1987) A model for the study of visceral pain states: chronic inflammation of the chronic decerebrate rat urinary bladder by irritant chemicals. *Pain* **28**, 109–27.
77. McMahon S.B., Dmitrieva N. and Koltzenburg M. (1995) Visceral pain. *Br J Anaesth* **75**, 132–44.
78. McNeill D.L. and Burden H.W. (1986) Convergence of sensory processes from the heart and left ulnar nerve onto a single afferent perikaryon: a neuroanatomical study in the rat employing fluorescent tracers. *Anat Rec* **214**, 441–4.
79. Menetrey D., Gannon A., Levine J.D. and Basbaum A.I. (1989) Expression of c-fos protein in interneurons and projection neurons of the rat spinal cord in response to noxious somatic, articular, and visceral stimulation. *J Comp Neurol* **285**, 177–95.
80. Merkord J. and Hennighausen G. (1989) Acute pancreatitis and bile duct lesions in rat induced by dibutyltin dichloride. *Exp Pathol* **36**, 59–62.
81. Merkord J., Jonas L., Weber H., Kroning G., Nizze H. and Hennighausen G. (1997) Acute interstitial pancreatitis in rats induced by dibutyltin dichloride (DBTC): pathogenesis and natural course of lesions. *Pancreas* **15**, 392–401.
82. Michaelis M., Häbler H.J. and Jäenig W. (1996) Silent afferents: a separate class of primary afferents? *Clin Exp Pharmacol Physiol* **23**, 99–105.
83. Minter R.M., Ferry M.A., Murday M.E., Tannahill C.L., Bahjat F.R., Oberholzer C., Oberholzer A., LaFace D., Hutchins B., Wen S., Shinoda J., Copeland E.M., 3 rd and Moldawer L.L. (2001) Adenoviral delivery of human and viral IL-10 in murine sepsis. *J Immunol* **167**, 1053–9.
84. Murphy A.Z., Suckow S.K., Johns M. and Traub R.J. (2009) Sex differences in the activation of the spinoparabrachial circuit by visceral pain. *Physiol Behav* **97(2)**, 205–12.
85. Nauta H.J., Hewitt E., Westlund K.N. and Willis W.D., Jr. (1997) Surgical interruption of a midline dorsal column visceral pain pathway. Case report and review of the literature. *J Neurosurg* **86**, 538–42.
86. Nauta H.J., Soukup V.M., Fabian R.H., Lin J.T., Grady J.J., Williams C.G., Campbell G.A., Westlund K.N. and Willis W.D., Jr. (2000) Punctate midline myelotomy for the relief of visceral cancer pain. *J Neurosurg* **92**, 125–30.
87. Ness T.J. and Gebhart G.F. (1991) Interactions between visceral and cutaneous nociception in the rat. II. Noxious visceral stimuli inhibit cutaneous nociceptive neurons and reflexes. *J Neurophysiol* **66**, 29–39.
88. Ness T.J. and Gebhart G.F. (1988) Colorectal distension as a noxious visceral stimulus: physiologic and pharmacologic characterization of pseudaffective reflexes in the rat. *Brain Res* **450**, 153–69.

89. Ness T.J. and Gebhart G.F. (2000) Acute inflammation differentially alters the activity of two classes of rat spinal visceral nociceptive neurons. *Neurosci Lett* **281**, 131–4.
90. Ness T.J., Richter H.E., Varner R.E. and Fillingim R.B. (1998) A psychophysical study of discomfort produced by repeated filling of the urinary bladder. *Pain* **76**, 61–9.
91. NIH (1998) NIH Guide: New Directions in Pain Research I. URL http://grants.nih.gov/grants/guide/pa-files/PA-98-102.html.
92. Noble M.D., Romac J., Vigna S.R. and Liddle R.A. (2008) A pH-sensitive, neurogenic pathway mediates disease severity in a model of post-ERCP pancreatitis. *Gut* **57**, 1566–71.
93. Olivar T., Cervero F. and Laird J.M. (2000) Responses of rat spinal neurones to natural and electrical stimulation of colonic afferents: effect of inflammation. *Brain Res* **866**, 168–77.
94. Ordway G.A., Longhurst J.C. and Mitchell J.H. (1983) Stimulation of pancreatic afferents reflexly activates the cardiovascular system in cats. *Am J Physiol* **245**, R820–6.
95. Ordway G.A., Mitchell J.H. and Longhurst J.C. (1982) Bradykinin stimulates pancreatic afferents to activate the cardiovascular system. *Trans Assoc Am Physicians* **95**, 229–36.
96. Palecek J., Paleckova V. and Willis W.D. (2003) Fos expression in spinothalamic and postsynaptic dorsal column neurons following noxious visceral and cutaneous stimuli. *Pain* **104**, 249–57.
97. Peters C.M., Lindsay T.H., Pomonis J.D., Luger N.M., Ghilardi J.R., Sevcik M.A. and Mantyh P.W. (2004) Endothelin and the tumorigenic component of bone cancer pain. *Neuroscience* **126**, 1043–52.
98. Pomonis J.D., Rogers S.D., Peters C.M., Ghilardi J.R. and Mantyh P.W. (2001) Expression and localization of endothelin receptors: implications for the involvement of peripheral glia in nociception. *J Neurosci* **21**, 999–1006.
99. Puig-Divi V., Molero X., Salas A., Guarner F., Guarner L. and Malagelada J.R. (1996) Induction of chronic pancreatic disease by trinitrobenzene sulfonic acid infusion into rat pancreatic ducts. *Pancreas* **13**, 417–24.
100. Puig-Divi V., Molero X., Vaquero E., Salas A., Guarner F. and Malagelada J. (1999) Ethanol feeding aggravates morphological and biochemical parameters in experimental chronic pancreatitis. *Digestion* **60**, 166–74.
101. Plusczyk T., Bersal B., Westermann S., Menger M. and Feifel G. (1999) ET-1 induces pancreatitis like microvascular deterioration and acinar cell injury. *J Surg Res* **85**, 301–10.
102. Plusczyk T., Witzel B., Menger M.D. and Schilling M. (2003) ETA and ETB receptor function in pancreatitis-associated microcirculatory failure, inflammation, and parenchymal injury. *Am J Physiol Gastrointest Liver Physiol* **285**, G145–53.
103. Qin C. and Foreman R.D. (2004) Viscerovisceral convergence of urinary bladder and colorectal inputs to lumbosacral spinal neurons in rats. *NeuroReport* **15**, 467–71.
104. Rack F.J. and Elkins C.W. (1950) Experiences with vagotomy and sympathectomy in the treatment of chronic recurrent pancreatitis. *AMA Arch Surg* **61**, 937–43.
105. Ray B.S. and Neill C.L. (1947) Abdominal visceral sensation in man. *Ann Surg* **126**, 709–24.
106. Ren Y., Zhang L., Lu Y., Yang H., Westlund K.N. (2009) Central lateral thalamic neurons receive noxious visceral mechanical and chemical input in rats. *J Neurophysiol* **102**, 244–58.
107. Rigamonti D.D. and Hancock M.B. (1974) Analysis of field potentials elicited in the dorsal column nuclei by splanchnic nerve A-beta afferents. *Brain Res* **77**, 326–9.
108. Rigamonti D.D. and Hancock M.B. (1978) Viscerosomatic convergence in the dorsal column nuclei of the cat. *Exp Neurol* **61**, 337–48.
109. Rodella L., Rezzani R., Gioia M., Tredici G. and Bianchi R. (1998) Expression of Fos immunoreactivity in the rat supraspinal regions following noxious visceral stimulation. *Brain Res Bull* **47**, 357–66.
110. Roza C., Laird J.M. and Cervero F. (1998) Spinal mechanisms underlying persistent pain and referred hyperalgesia in rats with an experimental ureteric stone. *J Neurophysiol* **79**, 1603–12.
111. Roza C. and Reeh P.W. (2001) Substance P, calcitonin gene related peptide and PGE2 co-released from the mouse colon: a new model to study nociceptive and inflammatory responses in viscera, in vitro. *Pain.* **93**, 213–9.
112. Ruch T.C. (1947) Visceral sensation and referred pain. In: *Howell's Textbook of Physiology*, 15th ed. (ed. by Fulton JF), pp. 385–401. Saunders, Philadelphia.
113. Ruda M.A. and Dubner R. (1992) Molecular and biochemical events mediate neuronal plasticity following inflammation and hyperalgesia. In: *Hyperalgesia and Allodynia* (ed. by Willis WD, Jr.), pp. 311–26. Raven, New York.

114. Saban M.R., Nguyen N.B., Hammond T.G. and Saban R. (2002) Gene expression profiling of mouse bladder inflammatory responses to LPS, substance P, and antigen-stimulation. *Am J Pathol* **160**, 2095–110.
115. Salzet M. (2001) Neuroimmunology of opioids from invertebrates to human. *Neuro Endocrinol Lett* **22**, 467–74.
116. Schaible H.G. and Schmidt R.F. (1983) Activation of groups III and IV sensory units in medial articular nerve by local mechanical stimulation of knee joint. *J Neurophysiol* **49**, 35–44.
117. Schmidt J., Compton C.C., Rattner D.W., Lewandrowski K. and Warshaw A.L. (1995) Late histopathologic changes and healing in an improved rodent model of acute necrotizing pancreatitis. *Digestion* **56**, 246–52.
118. Schmidt J., Rattner D.W., Lewandrowski K., Compton C.C., Mandavilli U., Knoefel W.T. and Warshaw A.L. (1992) A better model of acute pancreatitis for evaluating therapy. *Ann Surg* **215**, 44–56.
119. Sengupta J.N. and Gebhart G.F. (1994) Characterization of mechanosensitive pelvic nerve afferent fibers innervating the colon of the rat. *J Neurophysiol* **71**, 2046–60.
120. Sevcik M.A., Jonas B.M., Lindsay T.H., Halvorson K.G., Ghilardi J.R., Kuskowski M.A., Mukherjee P., Maggio J.E. and Mantyh P.W. (2006) Endogenous opioids inhibit early-stage pancreatic pain in a mouse model of pancreatic cancer. *Gastroenterology* **131**, 900–10.
121. Sharkey K.A., Williams R.G. and Dockray G.J. (1984) Sensory substance P innervation of the stomach and pancreas. Demonstration of capsaicin-sensitive sensory neurons in the rat by combined immunohistochemistry and retrograde tracing. *Gastroenterology* **87**, 914–21.
122. Sinclair D., Wedell H. and Feindel W. (1948) Referred pain and associated phenomena. *Brain* **71**, 184–211.
123. Sparmann G., Merkord J., Jaschke A., Nizze H., Jonas L., Lohr M., Liebe S. and Emmrich J. (1997) Pancreatic fibrosis in experimental pancreatitis induced by dibutyltin dichloride. *Gastroenterology* **112**, 1664–72.
124. Stein C. (1995) The control of pain in peripheral tissue by opioids. *N Engl J Med* **332**, 1685–90.
125. Steiner I., Spivack J.G., Deshmane S.L., Ace C.I., Preston C.M. and Fraser N.W. (1990) A herpes simplex virus type 1 mutant containing a nontransinducing Vmw65 protein establishes latent infection in vivo in the absence of viral replication and reactivates efficiently from explanted trigeminal ganglia. *J Virol* **64**, 1630–8.
126. Stone H.H. and Chauvin E.J. (1990) Pancreatic denervation for pain relief in chronic alcohol associated pancreatitis. *Br J Surg* **77**, 303–5.
127. Sugiura Y., Terui N. and Hosoya Y. (1989) Difference in distribution of central terminals between visceral and somatic unmyelinated (C) primary afferent fibers. *J Neurophysiol* **62**, 834–40.
128. Taylor B.K., Peterson M.A., Roderick R.E., Tate J., Green P.G., Levine J.O. and Basbaum A.I. (2000) Opioid inhibition of formalin-induced changes in plasma extravasation and local blood flow in rats. *Pain* **84**, 263–70.
129. Traub R.J. (2000) Evidence for thoracolumbar spinal cord processing of inflammatory, but not acute colonic pain. *Neuroreport* **11**, 2113–6.
130. Traub R.J., Pechman P., Iadarola M.J. and Gebhart G.F. (1992) Fos-like proteins in the lumbosacral spinal cord following noxious and non-noxious colorectal distention in the rat. *Pain* **49**, 393–403.
131. Traub R.J., Silva E., Gebhart G.F. and Solodkin A. (1996) Noxious colorectal distention induced-c-Fos protein in limbic brain structures in the rat. *Neurosci Lett* **215**, 165–8.
132. Ustinova E.E., Fraser M.O. and Pezzone M.A. (2006) Colonic irritation in the rat sensitizes urinary bladder afferents to mechanical and chemical stimuli: an afferent origin of pelvic organ cross-sensitization. *Am J Physiol Renal Physiol* **290**, F1478–87.
133. Vera-Portocarrero L.P., Lu Y. and Westlund K.N. (2003) Nociception in persistent pancreatitis in rats: effects of morphine and neuropeptide alterations. *Anesthesiology* **98**, 474–84.
134. Vizzard M.A. (2000) Increased expression of spinal cord Fos protein induced by bladder stimulation after spinal cord injury. *Am J Physiol Regul Integr Comp Physiol* **279**, R295–305.
135. Walker J.S. (2003) Anti-inflammatory effects of opioids. *Adv Exp Med Biol* **521**, 148–60.
136. Wang C.C. and Westlund K.N. (2001) Responses of rat dorsal column neurons to pancreatic nociceptive stimulation. *Neuro-Report* **12**, 2527–30.
137. Wang C.C., Willis W.D. and Westlund K.N. (1999) Ascending projections from the area around the spinal cord central canal: a Phaseolus vulgaris leucoagglutinin study in rats. *J Comp Neurol* **415**, 341–67.
138. Wesselmann U., Czakanski P.P., Affaitati G. and Giamberardino M.A. (1998) Uterine inflammation as a noxious visceral stimulus: behavioral characterization in the rat. *Neurosci Lett* **246**, 73–6.

139. Westlund K.N. (2009) Gene therapy for pancreatitis pain. *Gene Ther* **16**, 483–92.

140. Westlund, K.N., Vera- Portocarrero L.P., Zhang L., Wei J., Quast M.J., Palmer S.N. and Cleeland C.S. (2009) fMRI of supraspinal areas after morphine and one week pancreatic inflammation. *NeuroImage* **44**, 23–34.

141. Wick E.C., Hoge S.G., Grahn S.W., Kim E., Divino L.A., Grady E.F., Bunnett N.W. and Kirkwood K.S. (2006) Transient receptor potential vanilloid 1, calcitonin gene-related peptide, and substance P mediate nociception in acute pancreatitis. *Am J Physiol Gastrointest Liver Physiol* **290**, G959–69.

142. Wilson S.P. and Yeomans D.C. (2000) Genetic therapy for pain management. *Curr Rev Pain* **4**, 445–50.

143. Wilson S.P. and Yeomans D.C. (2002) Virally mediated delivery of enkephalin and other neuropeptide transgenes in experimental pain models. *Ann N Y Acad Sci* **971**, 515–21.

144. Wilson S.P., Yeomans D.C., Bender M.A., Lu Y., Goins W.F. and Glorioso J.C. (1999) Antihyperalgesic effects of infection with a preproenkephalin-encoding herpes virus. *Proc Natl Acad Sci USA* **96**, 3211–6.

145. Winfree C.J., Coombs D.W., DeLeo J.A. and Colburn R.W. (1992) Analgesic effects of intrathecally-administered 3 alpha-hydroxy-5 alpha-pregnan-20-one in a rat mechanical visceral pain model. *Life Sci* **50**,1007–12.

146. Wolfe D., Wechuck J., Krisky D., Mata M. and Fink D.J. (2009) A clinical trial of gene therapy for chronic pain. *Pain Med* **10**, 1325–30.

147. Wu T.T., Su Y.H., Block T.M. and Taylor J.M. (1998) Atypical splicing of the latency-associated transcripts of herpes simplex type 1. *Virology* **243**, 140–9.

148. Yang H., McNearney T.A., Chu R., Lu Y., Ren Y., Yeomans D.C., Wilson S.P. and Westlund K.N. (2008) Enkephalin-encoding herpes simplex virus-1 decreases inflammation and hotplate sensitivity in a chronic pancreatitis model. *Mol Pain* **4**, 8.

149. Yoshimura N. and de Groat W.C. (1999) Increased excitability of afferent neurons innervating rat urinary bladder after chronic bladder inflammation. *J Neurosci* **19**, 4644–53.

150. Yotsumoto F., Manabe T. and Ohshio G. (1993) Bradykinin involvement in the aggravation of acute pancreatitis in rabbits. *Digestion* **54**, 224–30.

151. Zhang H.Q., Rong P.J., Zhang S.P., Al-Chaer E.D. and Willis W.D. (2003) Noxious visceral inputs enhance cutaneous tactile response in rat thalamus. *Neurosci Lett* **336**, 109–12.

152. Zhou L., Zhang Q., Stein C. and Schafer M. (1998) Contribution of opioid receptors on primary afferent versus sympathetic neurons to peripheral opioid analgesia. *J Pharmacol Exp Ther* **286**, 1000–6.

Chapter 4

Animal Models of Pain After Peripheral Nerve Injury

Lintao Qu and Chao Ma

Abstract

Chronic pain can originate from injury or dysfunction in the peripheral nervous system, and currently available therapeutic methods are often ineffective. Despite the clinical significance, the mechanisms underlying the development and maintenance of chronic pain after peripheral nerve injury are obscure. During the last three decades, a number of animal models have been developed to study the chronic pain after peripheral nerve injury. Some of these animal models have been widely used in both academic research and pharmaceutical industry. The employment of animal models has greatly promoted our knowledge of the underlying mechanisms, suggested and tested new treatment strategies for chronic pain. This chapter reviewed the most important and widely used animal models of pain after peripheral nerve injury. Each of these animal models can produce a unique set of pain-related behavioral changes that are analogous to specific human chronic pain conditions. Improvements in the design and assessment of animal models will continue facilitating the research and treatment of chronic pain.

1. Introduction

Chronic pain is one of the most frequently reported clinical symptoms, affecting approximately one sixth of the population (1). Current therapeutics for chronic pain are often unsatisfactory, largely due to the lack of understanding of the underlying mechanisms. Injury, inflammation or dysfunction in the peripheral nervous system are common causes of chronic pain in humans (1–3), although other factors such as inflammation or lesions in brain or spinal cord and other diseases like cancer may also be involved (4–9). However, the mechanisms by which peripheral nerve injury produces chronic pain are poorly understood.

Chronic pain can be generally classified into two major categories: neuropathic pain and inflammatory pain, although the underlying mechanisms are always intertwined (3). Neuropathic pain refers to pain that originates from a primary lesion or dysfunction in the nervous system (1–3). According to the original location of the injury or dysfunction, neuropathic pain can be

Chao Ma and Jun-Ming Zhang (eds.), *Animal Models of Pain*, Neuromethods, vol. 49,
DOI 10.1007/978-1-60761-880-5_4, © Springer Science+Business Media, LLC 2011

further classified as peripheral or central. However, the peripheral or central site of underlying pathophysiology can no longer be used as a discriminandum, because both peripheral and central mechanisms are often involved in the development of pain, especially in peripheral neuropathic pain (1, 3). The pain syndrome caused by peripheral nerve injury is usually termed as neuropathic pain, which is characterized by spontaneous pain combined with hyperalgesia and allodynia (10, 11).

Although an abundance of human studies have focused on chronic pain evoked by peripheral nerve injury, the key issues relating to the pathogenesis of chronic pain cannot be resolved completely in human studies. For example, the evaluation of chronic pain in humans is complex because only stimuli that do not produce irreversible harm can be used in these subjects. In addition, human studies are limited by the inability to analyze the potential cellular and molecular mechanisms of chronic pain due to ethical issues. Therefore, there is a need to use animal models of periphery nerve injury to address these issues (12, 13). During the last 30 years, numerous animal models have been developed to simulate specific human chronic pain conditions and provided insights into the mechanisms of chronic pain (1–3, 8, 9). A better understanding of the underlying mechanisms of chronic pain might lead to new strategies of prevention or treatment.

Rodents are the most commonly used animal for chronic pain models, although primates (14) and other mammals have been used. Because pain is difficult if not impossible to measure directly in animals, it must be inferred from the measurement of pain-related behaviors, including (but not limited to) the changes in general behaviors, spontaneous pain behaviors (15–18), and responses evoked by various types of mechanical and thermal stimuli (1–3, 9) as well as the pain-inducing chemical agents such as formalin (19) or capsaicin (20). Because of the high variability in animal behavioral data due to a number of factors such as the individual differences of animals and subtle modifications in surgical procedures and behavioral testing protocols, one should be cautious when comparing and interpreting the results from different literatures.

2. Animal Models of Pain After Peripheral Nerve Injury

Although the neuropathic pain symptoms have been reported for over 150 years (21), there was no proper animal model until about 30 years ago, when Wall et al. first established the neuroma model by completely transecting the sciatic nerve (15). However, the neuroma model produced signs of deafferentation (such as autotomy, see Note 1) instead of pain. It was later learnt that the neuropathic pain model can be produced successfully by injury applied to part of a major nerve – such as chronic constriction injury (CCI) (22) and partial sciatic ligation (PSL) model (23) – or its branches – such as spinal nerve ligation (SNL) (24) and spared nerve injury (SNI) model (25). The intact nerve fibers which are adjacent to the injured/axotomized ones are believed to play a critical role in the development of evoked pain symptoms such as hyperalgesia and allodynia (26–28). In addition, it was discovered that neuropathic pain could be induced after injury to the dorsal root ganglion (DRG) (29, 30), dorsal root (26, 27, 31), or even the ventral root (32, 33). It was also found that axotomy may not be a necessity for neuropathic pain, since chronic compression of lumbar DRGs could produce similar painful symptoms without axotomy (29, 30, 34, 35). A summary diagram for animal models of neuropathic pain was given in Fig. 1.

2.1. Animal Models of Pain Induced by Complete Axotomy of One or More Major Nerve

The neuroma or axotomy-autotomy model was developed by Wall et al. in 1979 (15) to study the clinical conditions resulted from complete nerve transection, e.g., after amputation. This model is produced by a complete transection (axotomy) of a major peripheral nerve of rats or mice, usually the sciatic nerve, often with an accompanying lesion to the saphenous nerve, to

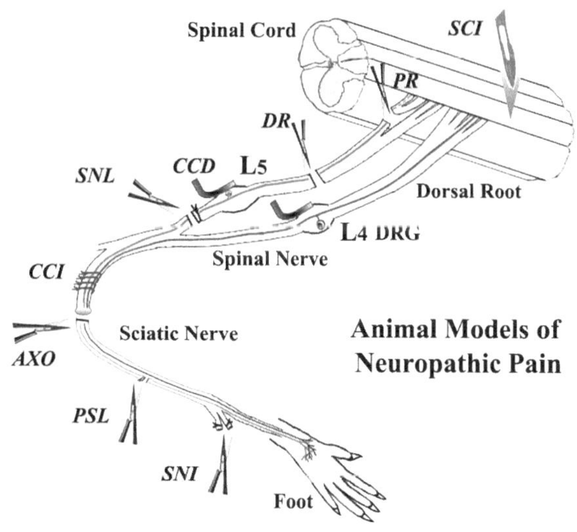

Fig. 1. Schematic drawing of some typical animal models of neuropathic pain. The abbreviations of animal models are listed in *shadowed boxes*, corresponding to different types and locations of injury to the peripheral or central nervous system. *SCI* spinal cord injury models, *PR* partial rhizotomy model, *DR* dorsal rhizotomy model, *CCD* chronic compression of DRG model, *SNL* spinal nerve ligation or Chung model, *CCI* chronic constriction injury to the sciatic nerve or Bennet–Xie model, *AXO* neuroma or axotomy–autotomy model, *PSL* partial sciatic nerve ligation or Seltzer model, *SNI* spared nerve injury model. See text for details (Reprint with permission from International Anesthesiology Clinics, Lippincott Williams and Wilkins (8)).

produce an entire denervation of the distal hindlimb (15). A neuroma is then formed in the proximal nerve stump after axotomy. The difficulty with such a model is that one has to rely on a spontaneous outcome measure, since no responses can be evoked by an applied stimulus. In this case, the term "autotomy" is used to describe the self-attack and mutilation of the anesthetic limb in animals after peripheral nerve injury (15, 36, 37) (see Note 1). The extent of self-mutilation is assessed by measuring the number and size of wounds on the denervated extremity (15). The autotomy score, defined on these criteria, is used to show the time course of abnormal behavior as well as the factors modulating it. In this model, the autotomy score continually increased during the observation time of 70 days, and the autotomy behavior could be exaggerated by encapsulation of the cut end of nerve in a polythene tube or by ligating both saphenous and sciatic nerves, whereas saphenous nerve transection alone did not produce autotomy (15). A sciatic nerve crush lesion caused only minimal attack (15) (see Note 1). The abnormal, ectopic discharges from both the neuroma and the cell bodies (somata) of injured neurons were found critical to the development and maintenance of pain-related behavioral changes (3, 38).

It is widely assumed that autotomy behavior following a nerve lesion is a sign of spontaneous pain (36), although it has been argued that autotomy is a result of excessive grooming in the absence of any sensory feedback (39–42). Both peripheral and central mechanisms have been proposed to contribute to autotomy (36, 40, 43), and genetic factors may also play a role (44). The axotomy-autotomy model was once the most widely used model for neuropathic pain before the introduction of partial nerve injury models, and was still used by some research groups nowadays despite of ethical concerns. Although this model to some extent mimics the spontaneous "phantom pain" in patients with amputation, most pain-related behaviors such as hyperalgesia and allodynia are absent due to the complete denervation of the distal hindlimb (15).

Methods other than mechanical force may be used to produce injury in a peripheral nerve. For example, in the sciatic cryoneurolysis model, peripheral mononeuropathy was produced by freezing the common sciatic nerve, a technique termed as sciatic cryoneurolysis (SCN) (45, 46). SCN could induce bilateral mechanical allodynia (but not thermal hyperalgesia) as well as autotomy (46).

2.2. Animal Models of Pain Induced by Partial Nerve Injury

2.2.1. The CCI Model

The chronic constriction injury (CCI) or Bennet–Xie model is the first animal model involving a partial injury of peripheral nerve that is analogous to simulate the clinical condition of chronic nerve compression such as the one that occurs in nerve entrapment neuropathy or spinal root irritation by a lumbar disk herniation (22).

First published by Bennet and Xie in 1988 (22), the CCI surgery involves loose ligation of the sciatic nerve of rat with four chromic gut sutures at the mid-thigh level to produce a CCI to the nerve without complete axotomy. After surgery, the animal exhibited heat and mechanically evoked hyperalgesia as well as cold and tactile allodynia, which lasted for a period of more than 2 months (13, 22, 47). Behavioral signs of spontaneous pain such as guarding posture, excessive licking and limping, and avoidance of weight bearing on the injured hind limb were also frequently observed, with rare cases of autotomy (13, 22, 47). In addition, animals with CCI showed the retardation of weight gain, which is probably a consequence of spontaneous pain. The degree and temporal course of the retardation of weight gain in the CCI model are similar to those in the neuroma model (22). Unlike the neuroma model where a peripheral nerve was completely transected, histological examination of the injured nerve in the CCI model revealed a partial denervation of the sciatic nerve that affected myelinated A-fibers while most unmyelinated C-fibers remained intact (22). The partial denervation of the sciatic nerve observed in the CCI model allows for the analysis of pain behaviors evoked by stimulation of the nerve's target – the hind paw. Due to the simplicity of surgical procedure and similarity to pain-related symptoms in humans, CCI model is still among the most widely used animal models of neuropathic pain in both basic research and pharmaceutical industry after over 20 years of invention. However, experiences of the surgical technique would be required while producing the CCI model as it is difficult to gage or reproduce the exact tension for the ligation needed (13, 22).

Aside from CCI in the sciatic nerve, a similar injury to other nerves has been applied to study neuropathic pain in different body regions. For example, CCI of the infraorbital nerve serves as a model of trigeminal pain (48).

2.2.2. The PSL Model

The partial sciatic nerve ligation (PSL) or Seltzer model was developed independently at about the same time as the CCI model, although the full description of the PSL was published 2 years later (23, 49). It is produced by unilateral tight ligation of one third to one half of the sciatic nerve at the upper-thigh level with a single ligature (23, 49). At this site, the sciatic nerve is not fasciculated into its main distal components. Thus, partial injury eliminated sensory fibers spread evenly across the innervations of the hind paw.

Starting hours after the operation and lasting for up to 7 months thereafter, rats receiving PSL developed tactile-defensive behavior of the partially denervated hind paw and often licked it, suggesting a sign of spontaneous pain (23). However, none of rats developed autotomy (23). In addition, bilateral touch-evoked allodynia and mechanical hyperalgesia, as well as ipsilateral thermal hyperalgesia and cold allodynia were observed in the nerve-injured rats (13, 23, 47, 50). Like the CCI model, the PSL model also involves a partial denervation of the sciatic nerve, but these two models produce some distinctively different behavioral disorders. In the PSL model, the rats showed a bilateral sharp decrease in withdraw thresholds to repetitive fine touch (mechanical hyperesthesia or allodynia) and von-Frey hair stimulation (mechanical hyperesthesia), which was mediated principally by myelinated A-fibers (23, 50). However, this feature did not appear in the CCI model. It might be because the low-threshold myelinated A-fibers in the undamaged portion of the nerve were blocked by the injury in the CCI model (22) but not in the PSL model (23, 50). Maintenance and production of the sensory disorders observed in the PSL model depended on the integrity of sympathetic outflow. Sympathectomy at the time of partial nerve injury aggravated the sensory disorders during the first few days, whereas these disorders were alleviated when sympathectomy was performed in rats several months after the nerve injury (51). However, the CCI model only showed a slight reduction of neuropathic pain behaviors 1 week after sympathectomy (47).

Partial nerve injury is the main cause of sympathetically maintained causalgiform pain disorders in humans (52). The PSL model shows many symptoms that characterize causalgia in humans, including the complex combination of rapid onset, allodynia to touch, hyperalgesia, mirror image phenomena, and sympathetic dependency. Therefore, PSL may serve as a reliable model for syndromes of the "mirror image" pain in certain human patients with complex regional pain syndrome (Type II) resulting from a partial nerve injury (23, 51, 52). However, the behavioral effects produced by PSL model may be highly variable possibly due to the differences in location and size of the ligation between individual animals, which makes this model difficult to standardize (13).

2.2.3. The SNL Model

The spinal nerve ligation (SNL) or Chung model was first introduced by Kim and Chung in 1992 (24). The SNL surgery involves a unilateral tight ligation (sometimes followed by transection) of the L5 or both L5 and L6 segmental spinal nerves distal to the DRG, leaving the L4 and L3 (saphenous nerve) component of the sciatic nerve intact (24). Like the CCI and PSL models, the SNL model produces a partial denervation of the sciatic nerve. However, the SNL model is distinct in that the injury is created at the level of the spinal nerve (L5 and L6), where the dorsal and ventral roots join distal to DRG, but proximal to the lumbar plexus where the spinal nerves sort themselves into the various peripheral nerves. Thus, the uninjured afferent axons of the sciatic nerve are accessible via the fourth lumbar root in the SNL model (49).

The SNL model produced allodynia and hyperalgesia quickly after surgery and last for at least 4 months, accompanied by behavioral signs of spontaneous pain (24, 53). The general pattern and time course of evoked pain behaviors (mechanical and cold allodynia) in the SNL model were similar to those seen in the CCI and PSL models (47). However, the magnitude of mechanical allodynia was larger in the SNL model than those in the other two models (47). The decrease of neuropathic pain behaviors after lumbar sympathectomy was most apparent in the L5 and L6 SNL injury model among all three models (47), suggesting that the pain behaviors are at least in part due to abnormality in the sympathetic nervous system. However, lumber sympathectomy failed to reverse or prevent the behavioral changes induced by the L5 SNL alone (54, 55).

The SNL model is one of the most widely used animal models of neuropathic pain, especially in basic research. In comparison with the CCI and PSL models, the surgical procedure in the SNL model requires the most extensive surgical exposure, which increases the likelihood of nonresponders and the chances for collateral tissue damage (13). However, the SNL model has the advantage of a consistent injury (complete axotomy) of an isolated nerve(s) exclusively associated with one or two DRGs and spinal segments. Therefore, it provides an "ideal" preparation for electrophysiological or histological studies targeting the DRG or spinal cord, making it easy to compare the injured (L5/6) and adjacent (L4) primary sensory neurons as well as other related cells in the DRG or corresponding spinal segments. Similar to the axotomized neurons in L5 DRG, neuronal hyperexcitability and ectopic discharges could be recorded from the unaxotomized neurons in the adjacent L4 DRG, and were found to be critical to the development of hyperalgesia in the SNL model (26–28).

2.2.4. The SNI Model

The spared nerve injury (SNI) model was firstly developed in rats by Decosterd and Woolf (25) and then validated in mice by Bouquin et al. (56). This model involves a lesion of two of the three terminal branches of the sciatic nerve (tibial and common peroneal nerves) sparing the remaining sural nerve. Like other peripheral nerve injury models, the SNI model also produces partial denervation of the sciatic nerve. Most importantly, the SNI model offers the advantage of a distinct anatomical distribution with an absence of co-mingling of injured and noninjured nerve fibers distal to the lesion such as the injured and noninjured nerves, and it permits behavioral testing of the noninjured skin territories adjacent to the denervated areas (25). After the SNI surgical procedure, all animals developed robust mechanical allodynia and hyperalgesia as well as the cold allodynia with an early onset (<24 h) and prolonged duration (>6 months) (25). These behavioral changes were observed in the ipsilateral sural and to a lesser extent in saphenous territories, but not in the contralateral hindpaw (25). Unlike the substantial effects on pain symptoms in the SNL model, sympathectomy has limited effects on SNI-induced pain sensitivity, attenuating cold allodynia but not mechanical allodynia and hyperalgesia in the SNI model (57, 58). Moreover, the sympathetic dependency of SNI-induced cold allodynia was developed with a delay of many weeks (57). Therefore, it has been suggested that the SNI model is not an appropriate model of sympathetically maintained mechanical allodynia and hyperalgesia but could be used to examine the mechanisms of cold allodynia associated with sympathetically maintained pain conditions (58).

In addition to the original SNI model, a variant SNI models have been established by Lee et al. by transecting the different combinations of the three branches of the sciatic nerve (59). They found that the tibial and sural nerves transection (TST) resulted in the largest amount of mechanical allodynia, cold allodynia, and spontaneous pain as compared to other types of combinations of three branches of sciatic nerves (59). Transecting sural and common peroneal nerve did not produce any neuropathic pain signs. Rats with transection of the tibial, sural, and common peroneal nerves showed moderate mechanical allodynia and cold allodynia and moderate to low spontaneous pain, which was similar to those observed in the original SNI model (13, 59). Injury to the tibial nerve alone, leaving the sural and common peroneal nerves, produced vigorous cold allodynia but moderate mechanical allodynia and spontaneous pain. It was also found that TST-induced mechanical allodynia and cold allodynia were not changed by sympathectomy (59, 60).

2.2.5. Other Animal Models of Neuropathic Pain

Aside from peripheral nerve injury, animal models of neuropathic pain could also be induced by injury to the DRG (29, 30), dorsal root (26, 27, 31), or even ventral root (33). It was also found that axotomy may not be a necessity for neuropathic pain, since chronic compression of lumbar DRGs could produce similar painful symptoms without axotomy (30). Injury to the spinal cord is another major cause of neuropathic pain (61, 62). A summary diagram for animal models of neuropathic pain is given in Fig. 1. These models will be introduced in the following chapters.

Peripheral neuropathy can accompany a variety of metabolic disorders like diabetes (see Chap. 9), autoimmune diseases such as Guillain–Barré syndrome (63–65), and viral infections such as human immunodeficiency virus (see Chap. 10) or varicella-zoster virus (66, 67). Local inflammation to the nerve or spinal ganglion is also a common cause of neuropathic pain (see Chap. 6). A number of animal models have been developed to study the mechanism of these conditions. Some of these models will be described in other chapters.

3. Conclusion

Employment of the proper animal model is critical to pain research, and is necessary in the development and evaluation of new therapeutic methods. As one of the common causes of chronic pain, injury to a peripheral nerve is among the earliest and most frequently used methods to produce animal models for pain research. During the last three decades, numerous animal models of pain after peripheral nerve injury have been developed and widely applied in both academic research and pharmaceutical industry. Each of these animal models can produce a unique set of pain-related behavioral changes that are analogous to specific human chronic pain conditions. Improvements in the design and assessment of animal models will continue to promote our understanding of the mechanisms underlying chronic pain as well as other pathological conditions in humans, therefore providing novel treatment strategies.

4. Note

1. Autotomy may occur after different types of deafferentation, such as sciatic nerve transection (15, 36), sciatic cryoneurolysis (45, 46), multiple dorsal rhizotomy (68), spared nerve ligation (69), and occasionally after loose ligation of the nerve (22, 69). A sciatic nerve crush injury could cause minimal to medium level of autotomy, depending on the methods of lesion (15, 69). Similar observations were also made in humans and nonhuman primates (70, 71). Strain differences in autotomy behaviors were observed in mice (72).

Acknowledgments

Supported by grants NS065091 (C.M.) from NIH/NINDS and Research Startup Fund (C.M.) from the Department of Anesthesiology, Yale University School of Medicine.

References

1. Campbell, J. N., and Meyer, R. A. (2006) Mechanisms of neuropathic pain, *Neuron* **52**, 77–92.
2. Xie, Y.-K. (2000) Mechanism of the chronic pain generation, *Chin Sci Bull* **5**, 775–783.
3. Devor, M. (1999) The pathophysiology of damaged peripheral nerves, in *Textbook of Pain* (Wall, P. D., and Medlzack, R., Eds.), pp 129–164, Churchill Livingstone, Edinburgh.

4. Foley, K. M. (2000) Controlling cancer pain, *Hosp Pract (Minneap)* **35**, 101–108, 111–102.
5. Ji, R. R. (2004) Peripheral and central mechanisms of inflammatory pain, with emphasis on MAP kinases, *Curr Drug Targets Inflamm Allergy* **3**, 299–303.
6. Lesage, P., and Portenoy, R. K. (1999) Trends in cancer pain management, *Cancer Control* **6**, 136–145.
7. Yezierski, R. P. (2005) Spinal cord injury: a model of central neuropathic pain, *Neurosignals* **14**, 182–193.
8. Ma, C. (2007) Animal models of pain, *Int Anesthesiol Clin* **45**, 121–131.
9. Wang, L. X., and Wang, Z. J. (2003) Animal and cellular models of chronic pain, *Adv Drug Deliv Rev* **55**, 949–965.
10. Bonica, J. J. (1990) Causalgia and other reflex sympathetic dystrophies, in *The Management of Pain* (Bonica, J. J., Ed.), pp. 220–243, Lea and Febieger, Philadelphia, PA.
11. Mitchell, S. W. (1972) *Injuries of Nerves and their Consequences*, JB Lippincott, Philadelphia, PA.
12. Casey, K. L., and Dubner, R. (1989) Animal models of chronic pain: scientific and ethical issues, *Pain* **38**, 249–252.
13. Dowdall, T., Robinson, I., and Meert, T. F. (2005) Comparison of five different rat models of peripheral nerve injury, *Pharmacol Biochem Behav* **80**, 93–108.
14. Carlton, S. M., Lekan, H. A., Kim, S. H., and Chung, J. M. (1994) Behavioral manifestations of an experimental model for peripheral neuropathy produced by spinal nerve ligation in the primate, *Pain* **56**, 155–166.
15. Wall, P. D., Devor, M., Inbal, R., Scadding, J. W., Schonfeld, D., Seltzer, Z., and Tomkiewicz, M. M. (1979) Autotomy following peripheral nerve lesions: experimental anaesthesia dolorosa, *Pain* **7**, 103–111.
16. Kauppila, T., Kontinen, V. K., and Pertovaara, A. (1998) Weight bearing of the limb as a confounding factor in assessment of mechanical allodynia in the rat, *Pain* **74**, 55–59.
17. Djouhri, L., Koutsikou, S., Fang, X., McMullan, S., and Lawson, S. N. (2006) Spontaneous pain, both neuropathic and inflammatory, is related to frequency of spontaneous firing in intact C-fiber nociceptors, *J Neurosci* **26**, 1281–1292.
18. Mills, C. D., Grady, J. J., and Hulsebosch, C. E. (2001) Changes in exploratory behavior as a measure of chronic central pain following spinal cord injury, *J Neurotrauma* **18**, 1091–1105.
19. Hunskaar, S., and Hole, K. (1987) The formalin test in mice: dissociation between inflammatory and non-inflammatory pain, *Pain* **30**, 103–114.
20. Baumann, T. K., Simone, D. A., Shain, C. N., and LaMotte, R. H. (1991) Neurogenic hyperalgesia: the search for the primary cutaneous afferent fibers that contribute to capsaicin-induced pain and hyperalgesia, *J Neurophysiol* **66**, 212–227.
21. Scadding, J. W. (2004) Treatment of neuropathic pain: historical aspects, *Pain Med* **5 Suppl 1**, S3–S8.
22. Bennett, G. J., and Xie, Y. K. (1988) A peripheral mononeuropathy in rat that produces disorders of pain sensation like those seen in man, *Pain* **33**, 87–107.
23. Seltzer, Z., Dubner, R., and Shir, Y. (1990) A novel behavioral model of neuropathic pain disorders produced in rats by partial sciatic nerve injury, *Pain* **43**, 205–218.
24. Kim, S. H., and Chung, J. M. (1992) An experimental model for peripheral neuropathy produced by segmental spinal nerve ligation in the rat, *Pain* **50**, 355–363.
25. Decosterd, I., and Woolf, C. J. (2000) Spared nerve injury: an animal model of persistent peripheral neuropathic pain, *Pain* **87**, 149–158.
26. Li, Y., Dorsi, M. J., Meyer, R. A., and Belzberg, A. J. (2000) Mechanical hyperalgesia after an L5 spinal nerve lesion in the rat is not dependent on input from injured nerve fibers, *Pain* **85**, 493–502.
27. Ma, C., Shu, Y., Zheng, Z., Chen, Y., Yao, H., Greenquist, K. W., White, F. A., and LaMotte, R. H. (2003) Similar electrophysiological changes in axotomized and neighboring intact dorsal root ganglion neurons, *J Neurophysiol* **89**, 1588–1602.
28. Wu, G., Ringkamp, M., Hartke, T. V., Murinson, B. B., Campbell, J. N., Griffin, J. W., and Meyer, R. A. (2001) Early onset of spontaneous activity in uninjured C-fiber nociceptors after injury to neighboring nerve fibers, *J Neurosci* **21**, RC140.
29. Hu, S. J., and Xing, J. L. (1998) An experimental model for chronic compression of dorsal root ganglion produced by intervertebral foramen stenosis in the rat, *Pain* **77**, 15–23.
30. Song, X. J., Hu, S. J., Greenquist, K. W., Zhang, J.-M., and LaMotte, R. H. (1999) Mechanical and thermal hyperalgesia and ectopic neuronal discharge after chronic compression of dorsal root ganglia, *J Neurophysiol* **82**, 3347–3358.
31. Eschenfelder, S., Habler, H. J., and Janig, W. (2000) Dorsal root section elicits signs of neuropathic pain rather than reversing them in rats with L5 spinal nerve injury, *Pain* **87**, 213–219.

32. Wu, G., Ringkamp, M., Murinson, B. B., Pogatzki, E. M., Hartke, T. V., Weerahandi, H. M., Campbell, J. N., Griffin, J. W., and Meyer, R. A. (2002) Degeneration of myelinated efferent fibers induces spontaneous activity in uninjured C-fiber afferents, *J Neurosci* **22**, 7746–7753.
33. Sheth, R. N., Dorsi, M. J., Li, Y., Murinson, B. B., Belzberg, A. J., Griffin, J. W., and Meyer, R. A. (2002) Mechanical hyperalgesia after an L5 ventral rhizotomy or an L5 ganglionectomy in the rat, *Pain* **96**, 63–72.
34. Zhang, J.-M., Song, X. J., and LaMotte, R. H. (1999) Enhanced excitability of sensory neurons in rats with cutaneous hyperalgesia produced by chronic compression of the dorsal root ganglion, *J Neurophysiol* **82**, 3359–3366.
35. Ma, C., and LaMotte, R. H. (2007) Multiple sites for generation of ectopic spontaneous activity in neurons of the chronically compressed dorsal root ganglion, *J Neurosci* **27**, 14059–14068.
36. Kauppila, T. (1998) Correlation between autotomy-behavior and current theories of neuropathic pain, *Neurosci Biobehav Rev* **23**, 111–129.
37. Wall, P. D., Scadding, J. W., and Tomkiewicz, M. M. (1979) The production and prevention of experimental anesthesia dolorosa, *Pain* **6**, 175–182.
38. Wall, P. D., and Devor, M. (1983) Sensory afferent impulses originate from dorsal root ganglia as well as from the periphery in normal and nerve injured rats, *Pain* **17**, 321–339.
39. Blumenkopf, B., and Lipman, J. J. (1991) Studies in autotomy: its pathophysiology and usefulness as a model of chronic pain, *Pain* **45**, 203–209.
40. Coderre, T. J., Grimes, R. W., and Melzack, R. (1986) Deafferentation and chronic pain in animals: an evaluation of evidence suggesting autotomy is related to pain, *Pain* **26**, 61–84.
41. Rodin, B. E., and Kruger, L. (1984) Deafferentation in animals as a model for the study of pain: an alternative hypothesis, *Brain Res* **319**, 213–228.
42. Sweet, W. H. (1981) Animal models of chronic pain: their possible validation from human experience with posterior rhizotomy and congenital analgesia (Part I of the second John J. Bonica lecture), *Pain* **10**, 275–295.
43. Saade, N. E., Ibrahim, M. Z., Atweh, S. F., and Jabbur, S. J. (1993) Explosive autotomy induced by simultaneous dorsal column lesion and limb denervation: a possible model for acute deafferentation pain, *Exp Neurol* **119**, 272–279.
44. Inbal, R., Devor, M., Tuchendler, O., and Lieblich, I. (1980) Autotomy following nerve injury: genetic factors in the development of chronic pain, *Pain* **9**, 327–337.
45. DeLeo, J. A., Coombs, D. W., Willenbring, S., Colburn, R. W., Fromm, C., Wagner, R., and Twitchell, B. B. (1994) Characterization of a neuropathic pain model: sciatic cryoneurolysis in the rat, *Pain* **56**, 9–16.
46. Willenbring, S., DeLeo, J. A., and Coombs, D. W. (1994) Differential behavioral outcomes in the sciatic cryoneurolysis model of neuropathic pain in rats, *Pain* **58**, 135–140.
47. Kim, K. J., Yoon, Y. W., and Chung, J. M. (1997) Comparison of three rodent neuropathic pain models, *Exp Brain Res* **113**, 200–206.
48. Vos, B. P., Strassman, A. M., and Maciewicz, R. J. (1994) Behavioral evidence of trigeminal neuropathic pain following chronic constriction injury to the rat's infraorbital nerve, *J Neurosci* **14**, 2708–2723.
49. Bennett, G. J., Chung, J. M., Honore, M., and Seltzer, Z. (2003) Models of neuropathic pain in the rat, *Curr Protoc Neurosci* **Chapter 9**, Unit 9, 14.
50. Shir, Y., and Seltzer, Z. (1990) A-fibers mediate mechanical hyperesthesia and allodynia and C-fibers mediate thermal hyperalgesia in a new model of causalgiform pain disorders in rats, *Neurosci Lett* **115**, 62–67.
51. Shir, Y., and Seltzer, Z. (1991) Effects of sympathectomy in a model of causalgiform pain produced by partial sciatic nerve injury in rats, *Pain* **45**, 309–320.
52. Seltzer, Z., and Shir, Y. (1991) Sympathetically-maintained causalgiform disorders in a model for neuropathic pain: a review, *J Basic Clin Physiol Pharmacol* **2**, 17–61.
53. Choi, Y., Yoon, Y. W., Na, H. S., Kim, S. H., and Chung, J. M. (1994) Behavioral signs of ongoing pain and cold allodynia in a rat model of neuropathic pain, *Pain* **59**, 369–376.
54. Ringkamp, M., Grethel, E. J., Choi, Y., Meyer, R. A., and Raja, S. N. (1999) Mechanical hyperalgesia after spinal nerve ligation in rat is not reversed by intraplantar or systemic administration of adrenergic antagonists, *Pain* **79**, 135–141.
55. Ringkamp, M., Eschenfelder, S., Grethel, E. J., Habler, H. J., Meyer, R. A., Janig, W., and Raja, S. N. (1999) Lumbar sympathectomy failed to reverse mechanical allodynia- and hyperalgesia-like behavior in rats with L5 spinal nerve injury, *Pain* **79**, 143–153.
56. Bourquin, A. F., Suveges, M., Pertin, M., Gilliard, N., Sardy, S., Davison, A. C., Spahn, D. R., and Decosterd, I. (2006) Assessment

and analysis of mechanical allodynia-like behavior induced by spared nerve injury (SNI) in the mouse, *Pain* **122**, 14e11–14e14.
57. Pertin, M., Allchorne, A. J., Beggah, A. T., Woolf, C. J., and Decosterd, I. (2007) Delayed sympathetic dependence in the spared nerve injury (SNI) model of neuropathic pain, *Mol Pain* **3**, 21.
58. Zhao, C., Chen, L., Tao, Y. X., Tall, J. M., Borzan, J., Ringkamp, M., Meyer, R. A., and Raja, S. N. (2007) Lumbar sympathectomy attenuates cold allodynia but not mechanical allodynia and hyperalgesia in rats with spared nerve injury, *J Pain* **8**, 931–937.
59. Lee, B. H., Won, R., Baik, E. J., Lee, S. H., and Moon, C. H. (2000) An animal model of neuropathic pain employing injury to the sciatic nerve branches, *Neuroreport* **11**, 657–661.
60. Han, D. W., Kweon, T. D., Kim, K. J., Lee, J. S., Chang, C. H., and Lee, Y. W. (2006) Does the tibial and sural nerve transection model represent sympathetically independent pain? *Yonsei Med J* **47**, 847–851.
61. Allen, A. R. (1911) Surgery of experimental lesion of spinal cord equivalent to crush injury of fracture dislocation of spinal column, *J Am Med Assoc* **57**, 878–880.
62. Balentine, J. D. (1978) Pathology of experimental spinal cord trauma. I. The necrotic lesion as a function of vascular injury, *Lab Invest* **39**, 236–253.
63. Hall, S. M., and Gregson, N. A. (1971) The in vivo and ultrastructural effects of injection of lysophosphatidyl choline into myelinated peripheral nerve fibres of the adult mouse, *J Cell Sci* **9**, 769–789.
64. Moulin, D. E., Hagen, N., Feasby, T. E., Amireh, R., and Hahn, A. (1997) Pain in Guillain–Barre syndrome, *Neurology* **48**, 328–331.
65. Wallace, V. C., Cottrell, D. F., Brophy, P. J., and Fleetwood-Walker, S. M. (2003) Focal lysolecithin-induced demyelination of peripheral afferents results in neuropathic pain behavior that is attenuated by cannabinoids, *J Neurosci* **23**, 3221–3233.
66. Fleetwood-Walker, S. M., Quinn, J. P., Wallace, C., Blackburn-Munro, G., Kelly, B. G., Fiskerstrand, C. E., Nash, A. A., and Dalziel, R. G. (1999) Behavioural changes in the rat following infection with varicella-zoster virus, *J Gen Virol* **80 (Pt 9)**, 2433–2436.
67. Sadzot-Delvaux, C., Merville-Louis, M. P., Delree, P., Marc, P., Piette, J., Moonen, G., and Rentier, B. (1990) An in vivo model of varicella-zoster virus latent infection of dorsal root ganglia, *J Neurosci Res* **26**, 83–89.
68. Wiesenfeld, Z., and Lindblom, U. (1980) Behavioral and electrophysiological effects of various types of peripheral nerve lesions in the rat: a comparison of possible models for chronic pain, *Pain* **8**, 285–298.
69. Casals-Diaz, L., Vivo, M., and Navarro, X. (2009) Nociceptive responses and spinal plastic changes of afferent C-fibers in three neuropathic pain models induced by sciatic nerve injury in the rat, *Exp Neurol* **217**, 84–95.
70. Levitt, M. (1985) Dysesthesias and self-mutilation in humans and subhumans: a review of clinical and experimental studies, *Brain Res* **357**, 247–290.
71. Pioli, E. Y., Gross, C. E., Meissner, W., Bioulac, B. H., and Bezard, E. (2003) The deafferented nonhuman primate is not a reliable model of intractable pain, *Neurol Res* **25**, 127–129.
72. Defrin, R., Zeitoun, I., and Urca, G. (1996) Strain differences in autotomy levels in mice: relation to spinal excitability, *Brain Res* **711**, 241–244.

Chapter 5

Animal Models of Pain After Injury to the Spinal Ganglia and Dorsal Roots

Xue-Jun Song

Abstract

Injury to dorsal root ganglion and/or the dorsal roots can lead to neuropathic pain. This chapter provides an overview of animal models that mimic ganglion compression by the implantation of a metal rod in the lumbar intervertebral foramen (CCD model) and dorsal root injury by loose ligation of the roots (DRC model) or partial dorsal rhizotomy (PDR model).

1. Introduction

Trauma, degenerative disorders, and other pathologies of the lumbar spine in humans can lead to chronic low back pain, sciatica, hyperalgesia, and other painful conditions of uncertain etiology. Peripheral nociceptors may be chronically activated in injured tissue, and injured primary afferent neurons that retain their central connections may have the development of ectopic discharges with potential nociceptive consequences. Conversely, a complete loss of afferent input from denervated tissue may give rise to pain of central origin (1). Animal models of painful sequelae in humans after nerve injury have provided behavioral evidence for ongoing pain and cutaneous hyperalgesia (2–4). Electrophysiological recordings from primary sensory neurons with transected peripheral axons indicate that the somata in the dorsal root ganglion (DRG) can become a source of both ectopic spontaneous discharges in the absence of peripheral receptor activation, and abnormal activity evoked by sympathetic stimulation and/or endogenous chemicals such as norepinephrine (5–8) and inflammatory mediators (9). Such abnormal activity, if occurring in the appropriate nociceptive afferent neurons, may maintain a state of

Chao Ma and Jun-Ming Zhang (eds.), *Animal Models of Pain*, Neuromethods, vol. 49,
DOI 10.1007/978-1-60761-880-5_5, © Springer Science+Business Media, LLC 2011

central sensitization of nociceptive neurons in the dorsal horn and, as a consequence, cause chronic pain and cutaneous hyperalgesia. Determination of the role of each functional class of afferents is problematic when axotomy has removed the injured neuron from its peripheral receptors. However, it is possible to do this in an animal model of neuropathic pain in which ectopic spontaneous discharges and other abnormal neuronal properties develop in neurons that do not undergo axotomy. Such abnormal properties may develop after spinal injuries or disorders which mechanically and/or chemically impinge on the DRG without affecting conduction in the spinal nerve or root.

This chapter provides an overview of a animal model of neuropathic pain after a chronic compression injury of the DRG (CCD model) produced by the implantation of a metal rod in the intervertebral foramen (IVF) (10–13). Injury to the dorsal root can also lead to neuropathic pain (12, 14–17). An overview of models of dorsal root constriction (DRC) (14, 16, 17) and partial dorsal rhizotomy (PDR) (12) in rats is also provided.

2. Materials

2.1. Animals

Adult, Sprague-Dawley rats (or other species) weighing 200–300 g (ranged 150–400 g) are used. Before and after surgery, the rats are housed in groups of four to five in plastic cages (~40 × 60 × 30 cm) with soft bedding and free access to food and water under a 12 h day/12 h night cycle.

2.2. Major Surgical Instruments and Materials

1. *Anesthesia:* balance (range 1–500 g); sodium pentobarbital or halothane; 2% procaine; syringe (2 ml).
2. *Sterilization:* dry sterilizer; betadine solution (10%); alcohol (70%); saline.
3. *Surgical instruments:* microscope (low power); light source provider; digital caliper; surgical gloves; heating pad; electric hair shaver; sterile stainless steel blades; blade handle; rongeurs; bone cutter; operating scissors; needle holder; blunt probe; alm retractor; supercut iris scissors; vannas scissors; tissue forceps; microdissection forceps; high and low temperature cautery; a sharp, stainless steel, L-shaped (4 mm in length, with an angle of ~80°) needle with a tip of approximately 0.4 mm diameter; injection syringe (1 ml); stainless steel rods, L-shaped, 4 mm in length and 0.4, 0.6, or 0.8 mm in diameter (see Note 1); silk sutures (4-0 and 7-0); surgical needles; glass handling forceps; glass needle with blunt tip (it can be easily made from the patch clamp capillary glass); cotton tip; and other supplies as needed.

3. Methods

3.1. CCD Model

Surgery is performed under anesthesia with sodium pentobarbital (40 mg/kg, i.p., supplemented as necessary) or other proper anesthetics. An incision is made between L_3 and L_6 approximately 0.5 cm away from the midline on one side. The paraspinal muscles on the side are bluntly separated from the mammillary process and the transverse process, and then the IVF of L_4 and/or L_5 exposed. A fine, sharp, stainless steel needle is inserted approximately 4 mm into IVF of L_4 and/or L_5, at a rostral direction (see Note 2). After the needle is withdrawn, a stainless steel rod, L-shaped, 4 mm in length, and 0.4–0.8 mm in diameter is implanted into the IVF at L_4 and/or L_5 ganglion (see Note 3). After the rod is in place, the muscle and skin layers are sutured. Relationships between the positions of the implanted rods and the DRG and lumbar vertebrae can be examined after behavioral testing and/or in vitro studies by performing a laminectomy and exposing the ganglia under anesthesia (see Note 4). Schematic of the CCD model is shown in Fig. 1.

3.2. DRC Model

The surgical procedure is illustrated in Fig. 2. The L_5–L_6 intervertebral space is identified by palpation, and a midline incision is made from L_1 to L_4. The paraspinal muscles are dissected free

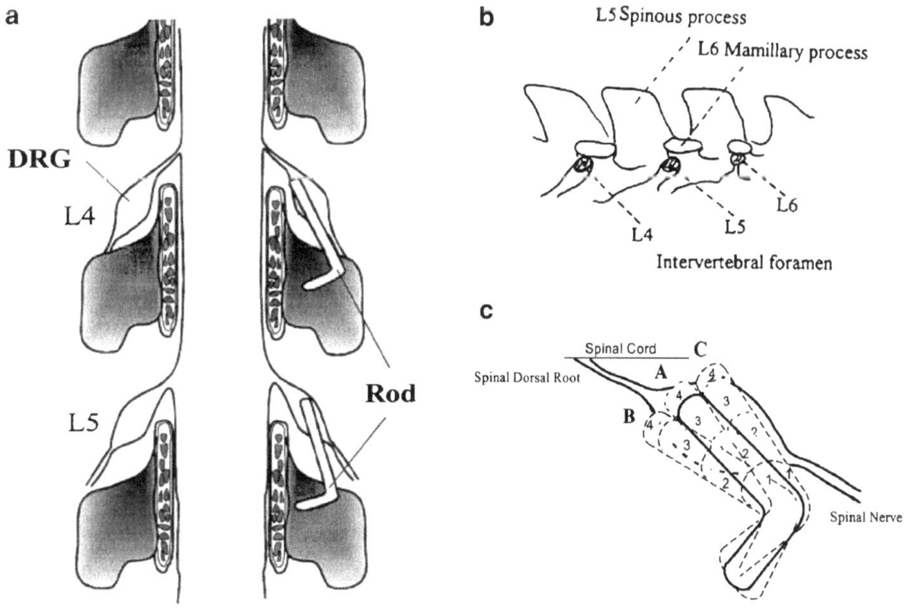

Fig. 1. Schematic of CCD model – intended relationship between the positions of the implanted rods and the DRG and lumbar vertebrae. L-shaped rods have been inserted ~4 mm into the L4 and L5 intervertebral foramen in a rostral direction at an angle of 30–40° ((**a**, **b**), modified from (10, 11)). (**c**) Relationship between the stainless steel rod and the compressed DRG at autopsy (11).

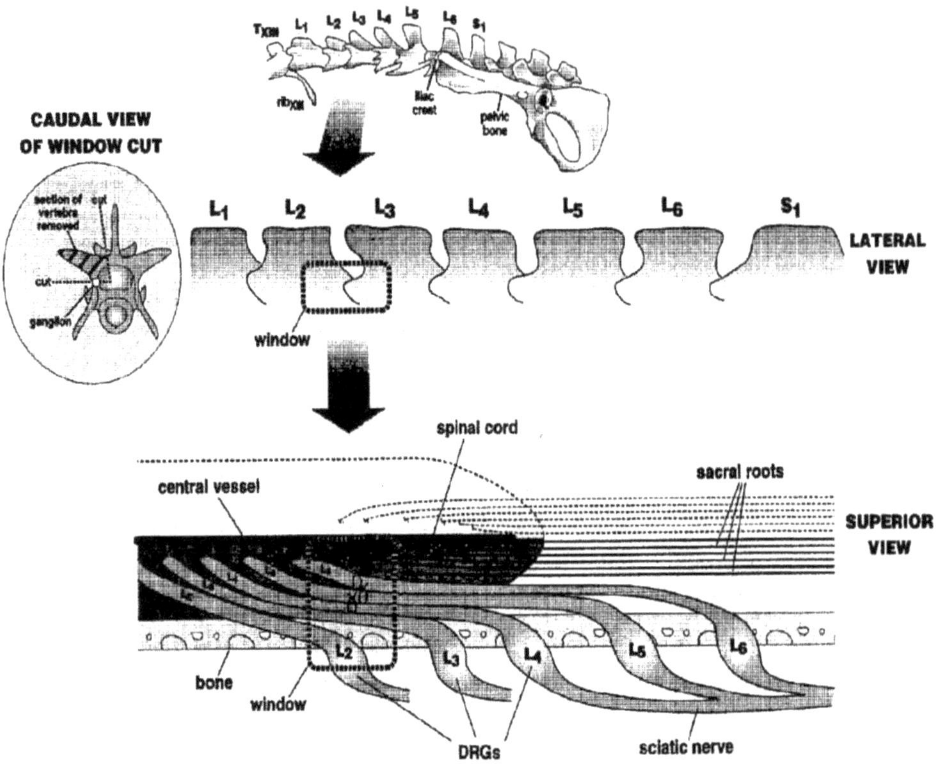

Fig. 2. Surgical procedure of making DRC model (14).

from the spinal processes on one side. The transverse processes of L_2 and L_3 are exposed by scraping off attached ligaments, and a laminectomy is performed to expose an approximately 1 cm length of spinal cord. The dura is opened over a 3–5 mm in length, and the L_4–L_6 dorsal roots are identified under a dissecting microscope. Either one or two 7-O silk ligatures are placed around each of the dorsal roots central to the DRG and tied using jeweler's forceps so that the dorsal roots are loosely constricted. Then the tissues, muscles, and the skin are closed with silk.

3.3. PDR Model

A midline incision is made from T_{12} to L_3. The paraspinal muscles are separated from the spinal processes on one side (usually the left). The transverse processes of L_1 and L_2 are exposed by scraping off attached ligaments, and a small "window" laminectomy is performed unilaterally at L_1–L_2 to expose L_4 and L_5 dorsal roots central to the DRGs and close to the spinal cord. The dura is opened ~5 mm in length, and the L_4 and L_5 dorsal roots are identified under a dissecting microscope (Fig. 2). The rootlets of L_4 and L_5 are carefully isolated from each other. Normally, there are four rootlets within each dorsal root. The rostral half of the rootlets (two out of four) are transected (~1–2 mm), whereas the caudal half is kept intact (Fig. 3; see Note 5).

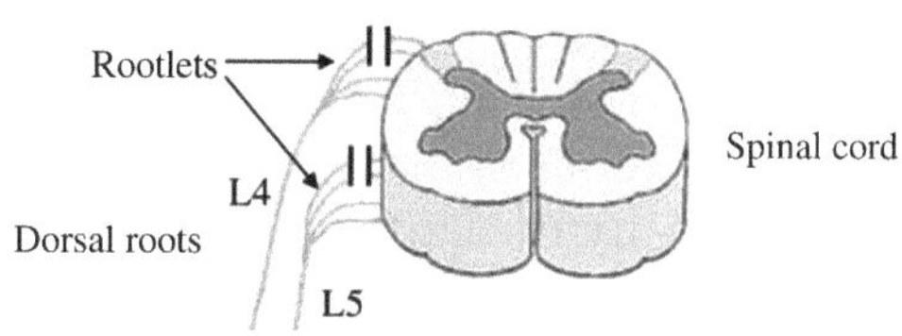

Fig. 3. Schematic of PDR model (modified from (12)).

4. Manifestations of Behavior and Sensory Neuron Electrophysiological Properties After CCD, DRC, or PDR Treatment

4.1. General Observations

The animals that received CCD, DRC, or PDR treatment appear in good health, but show a transient, small reduction in body weight in 1–2 postoperative weeks followed by resumption in weight gain. All rats developed varying degrees of abnormality in gait and posture after surgery. The rat exhibits some ataxia while walking within the first 15–24 h after surgery and thereafter exhibits some guarding of the paw ipsilateral to injury. In general, the affected hind paw is placed clumsily while walking and the toes, which are preoperatively spread apart while walking or standing, are together but not ventroflexed. The hind paw is everted and the animal stands and walks with the medial edge of the hind paw in contact with the floor. This is notable within the first 2 weeks after surgery but is less so during subsequent weeks. All rats walked normally and with toes spread apart in normal fashion when momentarily escaping a prod by the experimenter, suggesting that no permanent motor deficit is responsible for the abnormal gait and posture. It seems that any abnormal posture and gait most likely served the purpose of preventing aversive sensory stimulation.

4.2. Thermal and Mechanical Hyperalgesia and Tactile Allodynia

CCD or PDR treatment causes significant thermal hyperalgesia, which is seen the first day after surgery. Peak hyperalgesia is seen in the first 4 weeks, then waned gradually. It is worthy of noting that the severity and duration of thermal hyperalgesia produced by PDR is significantly less (~50%) and shorter (PDR vs. CCD: 4–5 vs. 9–11 weeks) than CCD (12). The hyperalgesia after CCD treatment recovers quickly after the rod withdrawal (13). Different from CCD and PDR, DRC treatment produces a transient increase in mean thermal paw withdrawal latency (hypoalgesia) with subsequent recovery (14). A rapid (<24 h for CCD and <3 days for DRC), long-lasting (7–10 weeks for CCD and ~22 weeks for DRC) mechanical and tactile allodynia is produced after CCD or DRC treatment (11–14).

4.3. Excitability and Plasticity of DRG and Spinal Dorsal Horn Neurons

All the three categories of DRG neurons, the large-, medium-, and small-sized neurons, become more excitable after CCD treatment. The increased excitability can be evaluated by increased incidence of spontaneous activity, depolarized membrane potential, decreased action potential threshold current, increased discharges following intracellular injection of depolarizing current, alterations of the potassium and sodium currents, and enhanced activity of signal transduction pathways (11, 12, 18–26). However, the PDR treatment does not significantly alter excitability of DRG neurons (12). Hyperexcitability of DRG neurons following nerve injury contributes to the sensitization of central nociceptive neurons in the dorsal horn, leading to chronic pain and hyperalgesia, which can be suppressed by intrathecal administration of NMDA receptor antagonists APV and MK-801, respectively (12). Dorsal root injury may have different effects on the development of hyperalgesia from DRG compression and peripheral nerve injury. These contributions involve different excitability mechanisms in the primary sensory neurons, but each involves pathways (presumably in the spinal cord) that depend upon NMDA receptors (12, 27).

5. Notes

1. The size of the rod (diameter) depends on the size of the rat. In general, the rods in 0.4, 0.6, and 0.8 mm in diameter are good for rats weighing 100–150, 150–300, and >300 g based on our observations.
2. A fine, sharp, stainless steel, L-shaped needle, ~0.4 mm in diameter with an angle (tip of the needle, ~4 mm in length, is bent to approximately 80°) to limit penetration is inserted approximately 4 mm into IVF of L_4 and/or L_5, at a rostral direction at an angle of about 30–40° to the dorsal middle line and −10 to −15° to the vertebral horizontal line. The purpose is to guide insertion of the rod into the IVF.
3. After the needle is withdrawn, a stainless steel rod, L-shaped, 4 mm in length and 0.4–0.8 mm in diameter is implanted into the IVF at L_4 and/or L_5 ganglion. Each insertion is guided by the mammillary process and the transverse process and oriented as described for the needle. As the rod is moved over the ganglion, the ipsilateral hind leg muscles typically exhibit one or two slight twitches. Depending on purpose of the specific experiment, one or two rods can be inserted into one or two IVF of L4 and/or L5. After the rod is in place, the muscle and skin layers are sutured.
4. Relationships between the positions of the implanted rods and the DRG and lumbar vertebrae. Intended relationships between the positions of the implanted rods and the DRG and lumbar vertebrae are shown in Fig. 1. As described above, the L-shaped rods are inserted into the IVF at L_4 and L_5 in a

rostral direction at an angle of 30–40° to the dorsal middle line and −10 to −15° to the vertebral horizontal line (Fig. 1a, b). The location of the implanted rod relative to the DRG can be further determined after termination of the experiment. A laminectomy is performed at the level of L4–L5. Each of the two rods, the two ganglia, and their dorsal roots and spinal nerves are identified and exposed. The position of each rod with respect to the DRG is examined under a dissection microscope and classified into one of the eight categories (Fig. 1c). Most of the rods (64 of 70 or 91.4% from 35 rats, 1–40 days after rod implantation) are in the positions of A2–4, B2–4, and C2–4, i.e., entirely or partly over the DRG. Six rods (8.6%) are in the positions of B4, or C4, indicating the possibility that the rod may have pressed against the dorsal root. However, there is no indication that the behavioral measurements of the hyperalgesia in rats with a deviant rod position are different in any way from those of the other animals. Especially note that the "L"-shaped, not the "l"-shaped, rod helps to prevent the rod falling into the spinal canal.

5. It is critical to identify correctly the targeted dorsal roots, L4 and/or L5 and to separate the rootlets. A local anesthetic such as procaine (2%) is dropped onto the rootlets 3–5 min prior to transection to avoid the rat jumps, which may damage the spinal cord severely.

References

1. Devor, M. (1994) The pathophysiology of damaged peripheral nerves. In: *Text Book of Pain*, edited by Wall PD, Melzack R. 3rd Ed. Churchill Livingstone, London, pp. 79–100.
2. Bennett, G.J. and Xie, Y.K. (1995) A peripheral mononeuropathy in rat that produces disorders of pain sensation like those seen in man. *Pain* **33**: 87–107.
3. Kim, S.H. and Chung, J.M. (1992) An experimental model for peripheral neuropathy produced by segmental spinal nerve ligation in the rat. *Pain* **50**: 355–63.
4. Seltzer, Z., Dubner, R., and Shir, Y. (1990) A novel behavioral model of neuropathic pain disorders produced in rats by partial sciatic nerve injury. *Pain* **43**: 205–18.
5. Wall, P.D. and Devor, M. (1983) Sensory afferent impulse originates from dorsal root ganglia as well as from the periphery in normal and nerve injury rats. *Pain* **17**: 321–39.
6. Xie, Y.K., Zhang, J.M., Petersen, M., and LaMotte, R.H. (1995) Functional changes in dorsal root ganglion cells after chronic nerve construction in the rat. *J Neurophysiol* **73**: 1811–20.
7. Zhang, J.M., Song, X.J., and LaMotte, R.H. (1997) An in vitro study of ectopic discharge generation and adrenergic sensitivity in the intact, nerve-injured rat dorsal root ganglion. *Pain* **72**: 51–7.
8. Hu, S.J., Yang, H.J., Jian, Z., Long, K.P., Duan, Y.B., Wan, Y.H., Xing, J.L., Xu, H., and Ju, G. (2000) Adrenergic sensitivity of neurons with non-periodic firing activity in rat injured dorsal root ganglion. *Neuroscience* **101**: 689–98.
9. Song, X.J., Zhang, J.M., and LaMotte, R.H. (2003) Somata of nerve-injured sensory neurons exhibit enhanced responses to inflammatory mediators. *Pain* **104**: 701–9.
10. Hu, S.J. and Xing, J. (1998) An experimental model for chronic compression of dorsal root ganglion produced by intervertebral foramen stenosis in the rat. *Pain* **77**: 15–23.
11. Song, X.J., Hu, S.J., Greenquist, K.W., Zhang, J.M., and LaMotte, R.H. (1999) Mechanical and thermal hyperalgesia and ectopic neuronal

discharge after chronic compression of dorsal root ganglia. *J Neurophysiol* **82**: 3347–58.

12. Song, X.J., Vizcarra, C., Xu, D.S., Rupert, R.L., and Wong, Z.N. (2003) Hyperalgesia and neural excitability following injuries to the peripheral and central branches of axon and somata of dorsal root ganglion neurons. *J Neurophysiol* **89**: 2185–93.
13. Song, X.J., Xu, D.S., Vizcarra, C., and Rupert, R.L. (2003) Onset and recovery of hyperalgesia and hyperexcitability of sensory neurons following intervertebral foramen volume reduction and restoration. *J Manipulative Physiol Ther* **26**: 426–36.
14. Tabo, E., Jinks, S.L., Eisele, J.H. Jr., and Carstens. E. (1999) Behavioral manifestations of neuropathic pain and mechanical allodynia, and changes in spinal dorsal horn neurons, following L4–L6 dorsal root constriction in rats. *Pain* **80**: 503–20.
15. Eschenfelder, S., Habler, H.J., and Janig, W. (2000) Dorsal root section elicits signs of neuropathic pain rather than reversing them in rats with L5 spinal nerve injury. *Pain* **87**: 213–9.
16. Kawakami, M., Weinstein, J.N., Chatani, K., Spratt, K.F., Meller, S.T., and Gebhart, G.F. (1994) Experimental lumbar radiculopathy. Behavioral and histologic changes in a model of radicular pain after spinal nerve root irritation with chromic gut ligatures in the rat. *Spine* **19**: 1795–802.
17. Kawakami, M., Weinstein, J.N., Spratt, K.F., Chatani, K., Traub, R.J., Meller, S.T., and Gebhart, G.F. (1994) Experimental lumbar radiculopathy. Immunohistochemical and quantitative demonstrations of pain induced by lumbar nerve root irritation of the rat. *Spine* **19**: 1780–94.
18. Song, X.J., Wang, Z.B., Gan, Q., and Walters, E.T. (2006) cAMP and cGMP pathways contribute to expression of hyperalgesia and sensory neuron hyperexcitability following dorsal root ganglion compression in the rat. *J Neurophysiol* **95**: 479–92.
19. Zhang, J.M., Song, X.J., and LaMotte, R.H. (1999) Enhanced excitability of sensory neurons in rats with cutaneous hyperalgesia produced by chronic compression of the dorsal root ganglion. *J Neurophysiol* **82**: 3359–66.
20. Hu, S.J., Song, X.J., and LaMotte, R.H. (2001) Protein kinase A mediates the generation of ectopic spontaneous activity in chronically compressed dorsal root ganglion. *Pain* **94**: 39–46.
21. Yao, H., Donnelly, D.F., Ma, C., and LaMotte, R.H. (2003) Upregulation of the hyperpolarization-activated cation current after chronic compression of the dorsal root ganglion. *J Neurosci* **23**: 2069–74.
22. Wang, Z.B., Gan, Q., Rupert, R.L., and Song, X.J. (2005) Thiamine, pyridoxine, cyanocobalamin and their combination inhibit thermal but not mechanical hyperalgesia in rats with primary sensory neuron injury. *Pain* **114**: 266–77.
23. White, F.A., Sun, J., Waters, S.M., Ma, C., Ren, D., Ripsch, M., Steflik, J., Cortright, D.N., Lamotte, R.H., and Miller, R.J. (2005) Excitatory monocyte chemoattractant protein-1 signaling is up-regulated in sensory neurons after chronic compression of the dorsal root ganglion. *Proc Natl Acad Sci U S A* **102**: 14092–7.
24. Ma, C., Greenquist, K.W., and LaMotte, R.H. (2006) Inflammatory mediators enhance the excitability of chronically compressed dorsal root ganglion neurons. *J Neurophysiol* **95**: 2098–107.
25. Ma, C. and LaMotte, R.H. (2007) Multiple sites for generation of ectopic spontaneous activity in neurons of the chronically compressed dorsal root ganglion. *J Neurosci* **27**: 14059–68.
26. Zheng, J.H., Walters, E.T., and Song, X.J. (2007) Dissociation of dorsal root ganglion neurons induces hyperexcitability that is maintained by increased responsiveness to cAMP and cGMP. *J Neurophysiol* **97**: 15–25.
27. Huang, Z.J. and Song, X.J. (2008) Differential alteration of sodium currents in nociceptive dorsal root ganglion neurons following injuries to the somata and peripheral branches of axons. *Mol Pain* **4**: 20.

Chapter 6

Localized Inflammatory Irritation of the Lumbar Ganglia: An Animal Model of Chemogenic Low Back Pain and Radiculopathy

Jun-Ming Zhang

Abstract

In this chapter, we describe a newly developed rodent model of chemogenic pain involving the inflammation of one or two lumbar sensory ganglia using the immune activator, zymosan. Using this model, we have investigated cellular, molecular, and ionic mechanisms of inflammatory responses within the dorsal root ganglion (DRG) and their contribution to the development of chemogenic pathological pain. DRG inflammation was induced by a single deposit of zymosan in the epidural space near the L5 DRG via a small hole drilled through the transverse process. After a single zymosan injection, rats developed bilateral mechanical hyperalgesia and allodynia which began by day 1 after surgery, peaked at days 3–7, and lasted up to 28 days. Robust satellite glial activation was observed in inflamed ganglia. Cytokine profile analysis using a multiplexing protein array system showed that local inflammatory irritation selectively increased proinflammatory cytokines/chemokines such as IL-1β, IL-6, IL-18, MCP-1, and GRO/KC up to 17-fold, and decreased anti-inflammatory cytokines such as IL-2 and IL-12 (p70) up to three-fold. Inflaming the DRG also remarkably increased the incidence of spontaneous activity of A- and C-fibers recorded in the dorsal root. Many of the spontaneously active A-fibers exhibited a short-bursting discharge pattern. Changes in cytokines and spontaneous activity correlated with the time course of pain behaviors, especially light stroke-evoked tactile allodynia. Local inflammation induced extensive sprouting of sympathetic fibers, extending from vascular processes within the inflamed DRG. Finally, patch clamp recording of the acutely dissociated DRG neurons revealed a significant upregulation of sodium and potassium channels. These results demonstrate the feasibility of inducing chronic localized inflammatory responses in the DRG in the absence of traumatic nerve damage and highlight the possible contribution of several inflammatory cytokines/chemokines to the generation of spontaneous activity and the development and persistence of chemogenic pathologic pain.

1. Introduction

Low back pain is one of the most frequent problems treated by orthopedic surgeons. Four out of five adults experience significant low back pain sometime during their lives. Aside from the common cold, problems caused by the lower back are the most

Chao Ma and Jun-Ming Zhang (eds.), *Animal Models of Pain*, Neuromethods, vol. 49,
DOI 10.1007/978-1-60761-880-5_6, © Springer Science+Business Media, LLC 2011

frequent cause of lost work days in adults under the age of 45 (1). Mechanical deformation and/or inflammatory irritation of the lumbar dorsal root ganglion (DRG) and its nerve roots is a possible consequence of certain disorders, such as spinal stenosis, disk herniation, spinal injury, or tumors (2–5). Clinical studies indicate that inflammation in the vicinity of the DRG may occur with or without apparent anatomic abnormalities. The mechanisms underlying inflammatory responses in the DRG and resultant low back pain and painful radiculopathy after compressive or noncompressive impact may be different. For example, spinal stenosis or spinal trauma-induced inflammation is related to acute or chronic mechanical compression and accompanied tissue damage, whereas in patients with disc rupture without herniation, the DRG and its nerve roots are exposed to nucleus pulposus, which is known to possess immunogenic capacities (6). An abnormally high concentration of phospholipase A_2, an inflammogen, was found in lumbar disc tissue at the affected level in low back pain patients (6). Cytokines, such as interleukins, have been detected in lumbar herniated disk materials in humans (7, 8). Evidence obtained from animal studies also suggests a critical role of the inflammatory response in the development of low back pain/radiculopathy. It was found that surgical exposure of the rat lumbar dorsal root and DRG subsequently reduced mechanical thresholds for ipsilateral foot withdrawal (9). In dog, an inflammatory response was produced in the spinal cord and in its roots after repetitive epidural injections of an extract from the nucleus pulposus (10). Cavanaugh et al. (11) reported that application of autologous nucleus pulposus to rabbit DRG in vitro evoked nerve discharge lasting 1–3 min. Recently, it has been discovered that a chronic implantation of a rod into the lumbar intervertebral foramen of rat, presumably compressing the DRG, produced behavioral evidence of chronic cutaneous hyperalgesia on the ipsilateral hind paw (12, 13). This is the first rodent model of low back pain/radiculopathy for the study of mechanical force in the cause of this clinical problem. Robust inflammatory responses have been demonstrated in the compressed DRGs as indicated by the increase in the number of satellite glia and macrophage (13). Cytokine profile analysis also revealed significantly elevated levels of key inflammatory cytokines/chemokines (14).

The response of sensory neurons to acute inflammatory mediators is well understood. However, little is known about how prolonged inflammation per se affects the DRG neurons, the changes in the cytokine profile within the DRG under pathological states, and the subsequent animal behavioral alterations. Numerous efforts and approaches have been engaged to examine inflammation or immune response as the cause of low back pain/radiculopathy. However, most studies are limited to the acute phase because of the lack of an animal model of chronic epidural or DRG inflammation.

In an attempt to examine inflammation per se in contributing to low back pain/radiculopathy, we have developed a new animal model of localized inflammation of the DRG (LID) (14). Using this model, we have examined possible causal factors that mediate pain following DRG inflammation. We characterized changes in cytokine production and their association with the development of spontaneous activity, as well as with pain and hyperalgesia. Changes in ion channels were examined to investigate the mechanisms underlying the generation of spontaneous activity of the inflamed DRG neurons. We have demonstrated the feasibility of the LID model in addressing neurological mechanisms of low back pain/radiculopathy and other pathologic states that involve inflammatory processes in the DRG.

2. Animal Model of Localized Inflammation of the Lumbar Ganglion

The LID model involves depositing a drop of the immune activator zymosan over one or two (L4 and L5) DRGs (14). Rats are anesthetized by continuous inhalation of isoflurane. On the right side, the paraspinal muscles are separated surgically from the L5–L6 vertebrae. The superior articular process of L6 and the transverse process of L5 vertebrae are cleaned. A small opening (diameter: 0.49 mm) is drilled through the junction of the transverse process and the lamina over the L5 DRG. A 25-G needle is cut short (length: 2–3 mm) and inserted into the opening without contacting the ganglion (15). The needle is connected to a microsyringe loaded with 20 μl zymosan suspended in incomplete Freund's adjuvant (IFA) at a concentration of 0.5 or 0.05 μg/μl corresponding to total amounts of 10 and 1 μg, respectively. The microsyringe is left in place for 2–3 min after injection of the 20 μl zymosan/IFA to prevent solution leakage. The sham group rats received a similar surgical procedure but without any injections.

3. General Observations of the Rats After Localized Inflammation of the DRG

All LID animals appear in good health throughout the testing period. The level of general activity is normal before and after surgery. Their fur is sleek and well groomed. No changes in gait or posture are observed in any rats of the sham or zymosan groups before or after surgery. There are no signs to indicate that rats are trying to reduce the weight placed upon the ipsilateral paw by leaning to the other side or by sitting on the opposite haunch, as observed in other rodent models of pathological pain. However, when a mechanical or thermal stimulus is applied to the hind paws during postoperative testing, the reflex withdrawals are of great amplitude and excessive duration during which the paw is held in the air typically 2–15 s. At times, a withdrawal of the hind paws is accompanied by exaggerated aversive behavior such as licking the stimulated paw. When the same stimuli are applied to the hind paw during preoperative testing, the reflex withdrawals are brief and of smaller amplitude.

4. Behavioral Changes in LID Rats

It was found that pain and hyperalgesia occurred in all rats after an injection of a small amount (1–10 μg) of the immune activator zymosan in IFA over one L5 DRG. The mechanical threshold to evoke a painful withdrawal response dropped to the lowest level within a week and remained low during the first two weeks. The tactile allodynia evidenced by painful foot withdrawal to light strokes using a piece of cotton wisp was present in all rats for the first two weeks. Although light stroke was not able to evoke withdrawal responses after POD14, cutaneous sensitivity remained higher than presurgical levels through the rest of our testing period. The time course of these pain behaviors correlated with the inflammatory responses in the DRGs, as indicated by elevated numbers of macrophages and satellite glia between days 1 and 14.

5. Changes in Cytokine Profiles in the Inflamed DRGs

Proinflammatory cytokines are often elevated in pain states (as well as in other pathologies of the nervous system) and often have anti-analgesic roles, though this cannot be uniformly assumed (16). Using rat cytokine protein array (or Luminex), we found that zymosan/IFA treatment significantly altered cytokine levels in the DRG. Consistent elevations, well above the levels seen in sham-operated animals, were observed for IL-1β, IL-6, IL-18, MCP-1, and GRO/KC cytokines that are traditionally classified as proinflammatory based on their effects on classical immune cells; in contrast, anti-inflammatory cytokines IL-2 and IL-12 levels declined. These results are consistent with those seen in DRG in some other pain models. Some of the elevated cytokines we observed, namely, IL-1β, IL-6, and MCP-1, have previously been found in the DRG in many studies; their receptors have shown to be present on neurons and/or satellite glia, and their levels increased in various pain models. Several lines of evidence indicate the importance of these cytokines in mediating chronic pain states: intrathecal injection and peripheral application of these cytokines can induce or enhance pain states; genetically modified mice lacking particular cytokines or their receptors have reduced pain in various models; these cytokines have acute and long-term excitatory effects on DRG neurons (including nociceptors) and can enhance release of pain neurotransmitters (for review, see (16, 17)).

One relatively new finding in the cytokine profile is the large increase in growth-related oncogene (GRO/KC also called CXCL1) in LID. There are few previous studies on the role of this cytokine in DRG, though it has been shown to enhance peptide release from neonatal DRG nociceptors in vitro and to induce hyperalgesia after peripheral injection (18). However, GRO/KC has been implicated in several CNS pathologies (e.g., (19–21)) and can acutely modulate activity of cerebellar neurons (22). Most recently, we have found that overnight GRO/KC incubation caused marked upregulation of Na^+ currents in acutely isolated small-diameter rat (adult) sensory neurons in vitro (23). In both the nonpeptidergic IB4-positive and peptidergic IB4-negative sensory neurons, TTX-resistant and TTX-sensitive currents increased two- to fourfold, without altered voltage dependence or kinetic changes. These effects required long exposures and were completely blocked by coincubation with the protein synthesis inhibitor cycloheximide. Amplification of cDNA from the neuronal cultures showed that three Na channel isoforms were predominant both before and after GRO/KC treatment (Nav 1.1, 1.7, and 1.8). TTX-sensitive isoforms 1.1 and 1.7 significantly increased two- to threefold after GRO/KC incubation, while 1.8 showed a trend towards increased expression. Current clamp experiments showed that GRO/KC caused a marked increase in excitability, including resting potential depolarization, decreased rheobase, and lower action potential threshold. Neurons acquired a striking ability to fire repetitively; IB4-positive cells also showed marked broadening of action potentials. This study suggests that GRO/KC may also have important pronociceptive effects via its direct actions on sensory neurons and may induce long-term changes that involve protein synthesis.

Similarly, there are few studies on the role of IL-18 in DRG, though it has been extensively studied in models of various CNS pathologies (24). In the LID model, IL-18 was notable for the more prolonged time course of its elevation. Another interesting result of the cytokine measurements was that there were no consistent increases in TNF-α levels in the inflamed DRG. In fact, on day 3, near the peak of pain behavior effects, we observed a small decrease in the level of TNF-α. In a previous study (25), it was found that perfusion of the compressed DRG with soluble TNF-α receptors led to a very small, albeit significant, decrease in pain behaviors. Thus, it suggests that TNF-α may play only a minor role in the pain induced by the DRG compression or localized inflammation (e.g., LID). This contrasts with the extensive evidence for a role of TNF-α in peripheral nerve injury and peripheral inflammation models. These results highlight the importance of examining a number of different cytokines, now that this has become technically feasible.

The sources of cytokines in the inflamed DRG are not yet known. Activated satellite glia and macrophages are known to produce a number of cytokines in the early phases of the inflammatory process and thus could contribute to the increased cytokine levels in the inflamed DRG. The slower time course of glial activation and macrophage infiltration compared with pain behaviors and changes in cytokine levels in the inflamed DRG might suggest that glia or macrophages are not likely to be the major sources of the elevated cytokines or a major cause of the pain behaviors. However, this interpretation is limited by the fact that these cells might rapidly release cytokines such as IL-1β, IL-18, IL-6, and TNF-α by converting them from the preexisting pools of the inactive forms into the active, releasable forms – such a release might occur much more rapidly than upregulation of glial activation marker proteins. Certain cytokines/chemokines such as MCP-1 and IL-1β can be synthesized and released from the sensory neurons in the DRG (26, 27). Another possible source of chemotactic cytokines is the endothelial cells, which can be activated by proinflammatory cytokines (28).

6. Generation of Spontaneous Activity in the Inflamed DRGs

The incidence of spontaneous activity in dorsal root fibers was high on POD 3–7 and dramatically decreased to a lower level by POD14. The time course for myelinated A-fibers (e.g., Aδ- and Aβ-) correlated well with the inflammatory response as well as the development of painful behaviors. In a recent in vitro study, it has been found that topical DRG application of an inflammatory cocktail consisting of histamine, prostaglandin E2 (PGE_2), serotonin (5-HT), and bradykinin robustly enhanced the excitability of the sensory neurons with intact or injured peripheral axons (29). In another study, MCP-1 was found to excite chronically compressed DRG neurons in vitro (26). These results and results from the LID model have led us to hypothesize that the initiation or development of abnormal spontaneous activity from inflamed DRG neurons may be attributed to the increased cytokine production and possibly to resultant elevated levels of other inflammatory mediators within the DRG. The incidence of spontaneous activity with a bursting pattern is significantly higher shortly after surgery, especially the short-burst discharges. Similar observations were made in chronically compressed DRGs (13). This suggests that in compressed DRGs, the bursting discharges may be caused by intraganglionic inflammation.

7. Local Inflammation Alters Excitability and Ion Currents in Small-Diameter Sensory Neurons

Animal models of chronic pain have generally been classified as either nerve injury or inflammation. Both nerve injury and inflammation can produce spontaneous pain, hyperalgesia, and allodynia (30). Both are associated with increased excitability and ectopic spontaneous discharges of DRG neurons (29, 31–34). These changes in sensory neurons, along with altered information processing in the spinal cord or higher centers, are thought to contribute to increased pain sensation (35). However, the underlying ionic mechanisms mediating the increased neuronal excitability are thought to differ between nerve injury and chronic inflammation. For example, nerve injury generally results in overall downregulation of Na^+ channels along with changes in the isoforms expressed, while chronic peripheral inflammation often results in upregulation of tetrodotoxin (TTX)-resistant Na^+ channels (36). Voltage-dependent K^+ currents are downregulated in almost all nerve injury and peripheral inflammation models, but the details differ between different models.

Recent work has suggested that the distinction between inflammation and nerve injury may not be complete. Nerve injury models often include some inflammation. For example, in peripheral nerve injury, macrophage infiltration, release of localized pro-inflammatory cytokines and their retrograde transport to the DRG have been shown to be important for hyperalgesia (37–39). Inflammatory processes and glial activation within the DRG have also been proposed to play important roles in some nerve injury models, though these processes have been more extensively studied in the central nervous system and peripheral nerve (for review (40–42)). To gain a better understanding of the inflammatory contributions to pathologic pain, patch clamp techniques were used on inflamed small-diameter DRG cells to investigate changes in Na^+ and K^+ channels that might contribute to the excitability changes observed in LID.

7.1. LID Increases Excitability of the DRG Neurons

The current clamp experiments demonstrated that several measures of excitability increased in small-diameter DRG neurons after localized inflammation. Hence, rheobase and action potential threshold were both reduced after LID. In addition, the rising and falling rates of the action potential increased. Increased excitability is a feature of virtually all animal models of chronic pain, whether they model nerve injury, peripheral inflammation, or compression of the DRG (43). It can be difficult to directly predict effects on excitability from voltage clamp measurements (44). However, the decreased rheobase and increased rising rate of the action potential seem at least qualitatively consistent with the observed increases in Na^+ currents (particularly TTX-sensitive currents, which activate more quickly and at more negative potentials after LID). It should be noted that other ion channels in sensory neurons may also play roles in chronic pain models, including the capsaicin receptor (45), hyperpolarization-activated cation channels (46), and voltage-gated calcium channels (47).

7.2. Voltage-Gated Sodium Channels

Changes in Na^+ channel expression or activity have been observed in other models of chronic pain. Nerve injury models generally lead to downregulation of TTX-R channels, with more complex effects on TTX-S channels. For example, after sciatic nerve transection, there is an upregulation of TTX-sensitive Na_v 1.3 channels and a downregulation of TTX-resistant Na^+ channels and of some other isoforms of TTX-sensitive channels. In this case, increased excitability is thought to result from redistribution of Na^+ channels along the axon and into the neuroma, as well as from the lower threshold and higher firing frequencies permitted by the switch to the Na_v 1.3 channel. Other nerve injury models such as spinal nerve ligation or chronic constriction injury also lead to reduced expression of TTX-resistant Na^+ channels, though the findings about TTX-sensitive currents are more variable (for review, see (36, 48)). In contrast, models of peripheral inflammation almost always lead to an upregulation of TTX-resistant current, although effects on the TTX-sensitive current are less consistent. The results from LID model are perhaps most similar to those seen in the subcutaneous carrageenan injection model, in which both TTX-sensitive and TTX-resistant currents (the latter due to increased Na_v 1.8 but not 1.9) are upregulated without shifts in voltage dependence of activation and inactivation (49, 50). Effects of peripheral inflammation on the TTX-S current are less consistent – particularly in models of visceral pain; there are examples of no change or even decreases of TTX-sensitive current after inflammation (51–53), though most of these studies do report increases of varying magnitude of the TTX-resistant current. The reasons for these differences between these various models are not yet clear. One possible explanation is that the visceral pain studies looked at a longer time point – 7–10 days instead of 3–5 days in our study and the carrageenan studies; perhaps, the effect on TTX-S current is transient. Another possibility is that our study examined both cutaneous and visceral afferent neurons which are mixed in the L4 and L5 DRG; a selective increase in TTX-S current within cutaneous neurons could account for this discrepancy.

The changes in Na^+ channels in nerve injury and inflammation models discussed above have, in general, been observed at the level of mRNA and protein, as well as in functional studies, and are generally thought to reflect changes in channel expression. However, posttranslational modifications and alternative splicing have also been proposed as additional possible mechanisms (48). These mechanisms are distinct from the rapid, reversible increase in Na^+ current caused by inflammatory mediators such as prostaglandins and bradykinin. Such effects, if present in the LID model in vivo, would not have been preserved in the acute culture methods used by this study and others cited above.

Small sensory neurons that are primarily nociceptors can be divided neurochemically into two populations: IB_4-positive, primarily nonpeptidergic neurons, and IB_4-negative peptidergic neurons. It has been shown that IB_4-positive neurons depend on glial-derived neurotrophic factor, whereas IB_4-negative neurons depend on nerve growth factor for survival during postnatal development (54). Furthermore, these two populations of nociceptors terminate in distinct regions of the superficial spinal cord. The finding that the IB4-negative cells had a higher expression of TTX-sensitive current than did IB4-positive cells is in general agreement with previous studies. In the normal DRG neurons, TTX-resistant currents with the characteristics of $Na_v1.8$ were commonly observed in both IB4-positive and IB4-negative cells, also in general agreement with previous electrophysiological and protein or mRNA expression studies (55–57). In LID, the increase in TTX-resistant current following DRG inflammation was confined to IB4-positive cells, suggesting that this may be an important variable to consider when conducting such studies.

7.3. Voltage-Gated K^+ Currents

A number of studies of voltage-gated K^+ current changes in acutely isolated, small DRG neurons after peripheral inflammation or nerve injury have been reported, in different laboratories and animal models. These studies differ markedly from the inflamed DRG neurons in that K^+ currents are increased in the LID, but are reduced in most studies involving either peripheral inflammation or nerve injury. A common finding is the functional reduction of fast-inactivating (transient) or sustained (delayed rectifier) components due to a reduced total conductance and/or to a leftward shift in steady-state inactivation that reduces channel availability at rest. This has been observed in several models of visceral inflammation, as well as in the axotomy model of nerve injury. In some but not all of these models, the sustained current is also reduced (31, 52, 58–63). In contrast, after LID, there is a 36% increase in the maximum magnitude (i.e., that evoked from a holding potential of −100 mV) of the sustained current, with comparable and significant increases at −60 and −50 mV, closer to the physiological resting potential. The fast-inactivating component increased in maximum amplitude after LID, also in contrast to most other models.

The excitability data indicate that the overall effect of the increases in both K^+ and Na^+ currents after LID is to increase excitability, which is in agreement with other models. A possible explanation for the differences in K^+ channel regulation between LID and other chronic pain models may lie in the magnitude of the Na^+ current changes observed. Although the increased Na^+ current density is qualitatively similar to the results obtained in many models of peripheral inflammation, the magnitude of the changes is generally much larger in the LID model. It may be that basic intrinsic mechanisms regulating overall excitability of the neuron (64) dictate an increase in K^+ currents to partially compensate for the very large increases in Na^+ currents. In addition, the LID model may expose cells to different types or higher concentrations of the regulatory molecules such as cytokines and growth factors which may regulate the ion channels (14). In this vein, it is interesting that a 24-h exposure to the proinflammatory cytokine IL-1β can increase TTX-sensitive Na^+ currents by 67% in trigeminal nociceptors (65).

8. Sympathetic Sprouting After Localized Inflammation

Sympathetic sprouting occurred in all inflamed DRGs beginning as early as day 3 after surgery. It can be clearly observed that sprouted fibers originate from the vascular processes in the DRGs. Fiber sprouting may be explained by increased expression of certain inflammatory cytokines such as IL-6 which have been reported to be able to cause sympathetic fiber growth (66, 67). Other factors such as NGF and NT-3 may be involved, too (68). It indicates that sympathetic sprouting may occur in intact DRG without peripheral axotomy. Extensive sympathetic sprouting in the inflamed DRG suggests a possible sympathetic component in this model, as reported in other inflammatory pain conditions (69).

9. Comparison Between LID and SIN Models

It is of interest to compare the LID model with the previously described sciatic inflammatory neuritis (SIN) model of Chacur et al. (70) in which pain was induced by applying zymosan around the sciatic nerve. Both models produce mechanical, but not thermal pain behaviors. SIN is associated with localized increases in IL-1β, as was observed in LID, and also results in increased TNF-α release. Both models produce both contralateral and ipsilateral pain at higher zymosan doses. Milligan et al. suggested that this phenomenon was possibly mediated by glial activation across the spinal cord. The bilateral allodynia induced by SIN can be prevented or reversed by intrathecal application of glial inhibitors, inhibitors of p38 mitogen-activated kinases, or proinflammatory cytokine antagonists specific for interleukin-1, TNF-α, or interleukin-6 (71). Contralateral allodynia and hyperalgesia in rats with zymosan/IFA treatment of a single DRG may share similar mechanisms.

10. Summary

Localized inflammation of the DRG results in a robust, prolonged state of mechanical hyperalgesia and allodynia. In contrast to the rapidly reversible effects of acute inflammatory mediators, the direct inflammatory irritation in the LID model leads to changes in neuronal properties that are preserved after isolation of the DRG from the animal and dilution or removal of soluble mediators. Isolated inflammation of the DRG, in the absence of nerve trauma or damage, occurs in certain clinical pain states such as postherpetic neuralgias and some forms of low back pain such as one following lumbar disc rupture. However, the LID model can also be viewed as one which allows to study the effects of DRG inflammation per se (i.e., in the absence of nerve damage). This is of more general interest since inflammation is an important component of most pain models including those based on nerve injury. Another advantage of this model, as for the DRG compression or CCD model, is that it allows in vivo localized perfusion of the DRG (25, 72), a great advantage for experimental manipulations.

Acknowledgements

This work was supported in part by NIH grants NS55860 and NS45594, and the University of Cincinnati Millennium Fund.

References

1. DeLeo, J. A., Colburn, R. W., Nichols, M., and Malhotra, A. (1996) Interleukin-6-mediated hyperalgesia/allodynia and increased spinal IL-6 expression in a rat mononeuropathy model, *J Interferon Cytokine Res 16*, 695–700.
2. Briggs, C. A. and Chandraraj, S. (1995) Variations in the lumbosacral ligament and associated changes in the lumbosacral region resulting in compression of the fifth dorsal root ganglion and spinal nerve, *Clin Anat 8*, 339–346.
3. Hoch, R. C., Rodriguez, R., Manning, T., Bishop, M., Mead, P., Shoemaker, W. C., and Abraham, E. (1993) Effects of accidental trauma on cytokine and endotoxin production, *Crit Care Med 21*, 839–845.
4. Poletti, C. E. (1996) Third cervical nerve root and ganglion compression: clinical syndrome, surgical anatomy, and pathological findings, *Neurosurgery 39*, 941–948; discussion 948–949.
5. Sato, K. and Kikuchi, S. (1993) An anatomic study of foraminal nerve root lesions in the lumbar spine, *Spine 18*, 2246–2251.
6. Saal, J. S., Franson, R. C., Dobrow, R., Saal, J. A., White, A. H., and Goldthwaite, N. (1990) High levels of inflammatory phospholipase A2 activity in lumbar disc herniations, *Spine 15*, 674–678.
7. Kang, J. D., Georgescu, H. I., McIntyre-Larkin, L., Stefanovic-Racic, M., and Evans, C. H. (1995) Herniated cervical intervertebral discs spontaneously produce matrix metalloproteinases, nitric oxide, interleukin-6, and prostaglandin E2, *Spine 20*, 2373–2378.
8. Yamagishi, M., Nemoto, O., and Kikuchi, T. (1992) Ruputured human disc tissues produce metalloproteinase and interleukin-1, *Trans Orthop Res Soc 17*, 190.

9. Olmarker, K. and Myers, R. R. (1998) Pathogenesis of sciatic pain: role of herniated nucleus pulposus and deformation of spinal nerve root and dorsal root ganglion, *Pain 78*, 99–105.
10. McCarron, R. F., Wimpee, M. W., Hudkins, P. G., and Laros, G. S. (1987) The inflammatory effect of nucleus pulposus. A possible element in the pathogenesis of low-back pain, *Spine 12*, 760–764.
11. Cavanaugh, J. M., Ozaktay, A. C., and Vaidyanathan, S. (1994) Mechano- and chemosensitivity of lumbar dorsal roots and dorsal root ganglia: an in vitro study, *Trans Orthop Res Soc 1994*, 109.
12. Hu, S. J. and Xing, J. L. (1998) An experimental model for chronic compression of dorsal root ganglion produced by intervertebral foramen stenosis in the rat, *Pain 77*, 15–23.
13. Song, X. J., Hu, S. J., Greenquist, K. W., Zhang, J.-M., and LaMotte, R. H. (1999) Mechanical and thermal hyperalgesia and ectopic neuronal discharge after chronic compression of dorsal root ganglia, *J Neurophysiol 82*, 3347–3358.
14. Xie, W. R., Deng, H., Li, H., Bowen, T. L., Strong, J. A., and Zhang, J.-M. (2006) Robust increase of cutaneous sensitivity, cytokine production and sympathetic sprouting in rats with localized inflammatory irritation of the spinal ganglia, *Neuroscience 142*, 809–822.
15. Zhang, J.-M., Homma, Y., Ackerman, W. E., and Brull, S. J. (2001) Topical application of acidic bupivacaine to the lumbar ganglion induces mechanical hyperalgesia in the rat, *Anesth Analg 93*, 466–471.
16. Watkins, L. R. and Maier, S. F. (2002) Beyond neurons: Evidence that immune and glial cells contribute to pathological pain states [Review], *Physiol Rev 82*, 981–1011.
17. Sommer, C. and Kress, M. (2004) Recent findings on how proinflammatory cytokines cause pain: peripheral mechanisms in inflammatory and neuropathic hyperalgesia, *Neurosci Lett 361*, 184–187.
18. Qin, X., Wan, Y., and Wang, X. (2005) CCL2 and CXCL1 trigger calcitonin gene-related peptide release by exciting primary nociceptive neurons, *J Neurosci Res 82*, 51–62.
19. Losy, J., Zaremba, J., and Skrobanski, P. (2005) CXCL1 (GRO-alpha) chemokine in acute ischaemic stroke patients, *Folia Neuropathol 43*, 97–102.
20. Valles, A., Grijpink-Ongering, L., de Bree, F. M., Tuinstra, T., and Ronken, E. (2006) Differential regulation of the CXCR2 chemokine network in rat brain trauma: Implications for neuroimmune interactions and neuronal survival, *Neurobiol Dis 22*, 312–322.
21. Xia, M. and Hyman, B. T. (2002) GROalpha/KC, a chemokine receptor CXCR2 ligand, can be a potent trigger for neuronal ERK1/2 and PI-3 kinase pathways and for tau hyperphosphorylation-a role in Alzheimer's disease? *J Neuroimmunol 122*, 55–64.
22. Giovannelli, A., Limatola, C., Ragozzino, D., Mileo, A. M., Ruggieri, A., Ciotti, M. T., Mercanti, D., Santoni, A., and Eusebi, F. (1998) CXC chemokines interleukin-8 (IL-8) and growth-related gene product alpha (GROalpha) modulate Purkinje neuron activity in mouse cerebellum, *J Neuroimmunol 92*, 122–132.
23. Wang, J. G., Strong, J. A., Xie, W., Yang, R. H., Coyle, D. E., Wick, D. M., Dorsey, E. D., and Zhang, J. M. (2008) The chemokine CXCL1/growth related oncogene increases sodium currents and neuronal excitability in small diameter sensory neurons, *Mol Pain 4*, 38.
24. Felderhoff-Mueser, U., Schmidt, O. I., Oberholzer, A., Buhrer, C., and Stahel, P. F. (2005) IL-18: a key player in neuroinflammation and neurodegeneration? *Trends Neurosci 28*, 487–493.
25. Homma, Y., Brull, S. J., and Zhang, J.-M. (2002) A comparison of chronic pain behavior following local application of tumor necrosis factor alpha to the normal and mechanically compressed lumbar ganglia in the rat, *Pain 95*, 235–246.
26. White, F. A., Sun, J., Waters, S. M., Ma, C., Ren, D., Ripsch, M., Steflik, J., Cortright, D. N., Lamotte, R. H., and Miller, R. J. (2005) Excitatory monocyte chemoattractant protein-1 signaling is up-regulated in sensory neurons after chronic compression of the dorsal root ganglion, *Proc Natl Acad Sci USA 102*, 14092–14097.
27. Copray, J. C., Mantingh, I., Brouwer, N., Biber, K., Kust, B. M., Liem, R. S., Huitinga, I., Tilders, F. J., Van Dam, A. M., and Boddeke, H. W. (2001) Expression of interleukin-1 beta in rat dorsal root ganglia, *J Neuroimmunol 118*, 203–211.
28. Martiney, J. A., Cuff, C., Litwak, M., Berman, J., and Brosnan, C. F. (1998) Cytokine-induced inflammation in the central nervous system revisited, *Neurochem Res 23*, 349–359.
29. Song, X. J., Zhang, J.-M., Hu, S. J., and LaMotte, R. H. (2003) Somata of nerve-injured sensory neurons exhibit enhanced responses to inflammatory mediators, *Pain 104*, 701–709.

30. Scholz, J. and Woolf, C. J. (2002) Can we conquer pain? *Nat Neurosci 5 Suppl*, 1062–1067.
31. Abdulla, F. A. and Smith, P. A. (2001) Axotomy- and autotomy-induced changes in Ca2+ and K+ channel currents of rat dorsal root ganglion neurons, *J Neurophysiol 85*, 644–658.
32. Kim, Y. I., Na, H. S., Kim, S. H., Han, H. C., Yoon, Y. W., Sung, B., Nam, H. J., Shin, S. L., and Hong, S. K. (1998) Cell type-specific changes of the membrane properties of peripherally-axotomized dorsal root ganglion neurons in a rat model of neuropathic pain, *Neuroscience 86*, 301–309.
33. Moore, B. A., Stewart, T. M., Hill, C., and Vanner, S. J. (2002) TNBS ileitis evokes hyperexcitability and changes in ionic membrane properties of nociceptive DRG neurons, *Am J Physiol Gastrointest Liver Physiol 282*, G1045–G1051.
34. Zhang, J.-M., Song, X. J., and LaMotte, R. H. (1999) Enhanced excitability of sensory neurons in rats with cutaneous hyperalgesia produced by chronic compression of the dorsal root ganglion, *J Neurophysiol 82*, 3359–3366.
35. Millan, M. J. (1999) The induction of pain: an integrative review, *Prog Neurobiol 57*, 1–164.
36. Waxman, S. G., Cummins, T. R., Dib-Hajj, S., Fjell, J., and Black, J. A. (1999) Sodium channels, excitability of primary sensory neurons, and the molecular basis of pain, *Muscle Nerve 22*, 1177–1187.
37. Taskinen, H. S. and Roytta, M. (2000) Increased expression of chemokines (MCP-1, MIP-1alpha, RANTES) after peripheral nerve transection, *J Peripher Nerv Syst 5*, 75–81.
38. Tanaka, T., Minami, M., Nakagawa, T., and Satoh, M. (2004) Enhanced production of monocyte chemoattractant protein-1 in the dorsal root ganglia in a rat model of neuropathic pain: possible involvement in the development of neuropathic pain, *Neurosci Res 48*, 463–469.
39. Kleinschnitz, C., Brinkhoff, J., Zelenka, M., Sommer, C., and Stoll, G. (2004) The extent of cytokine induction in peripheral nerve lesions depends on the mode of injury and NMDA receptor signaling, *J Neuroimmunol 149*, 77–83.
40. Ji, R. R. and Strichartz, G. (2004) Cell signaling and the genesis of neuropathic pain, *Sci STKE 2004*, reE14.
41. Marchand, F., Perretti, M., and McMahon, S. B. (2005) Role of the immune system in chronic pain, *Nat Rev Neurosci 6*, 521–532.
42. Moalem, G. and Tracey, D. J. (2006) Immune and inflammatory mechanisms in neuropathic pain, *Brain Res Brain Res Rev 51*, 240–264.
43. Devor, M. (2006) Response of nerves to injury in relation to neuropathic pain, in *Textbook of Pain* (McMahon, S. B. and Koltzenburg, M., Eds.) 5th ed., pp. 905–927, Elsevier/Churchill Livingstone, Philadelphia, PA.
44. Schild, J. H. and Kunze, D. L. (1997) Experimental and modeling study of Na+ current heterogeneity in rat nodose neurons and its impact on neuronal discharge, *J Neurophysiol 78*, 3198–3209.
45. Krause, J. E., Chenard, B. L., and Cortright, D. N. (2005) Transient receptor potential ion channels as targets for the discovery of pain therapeutics, *Curr Opin Investig Drugs 6*, 48–57.
46. Chaplan, S. R., Guo, H. Q., Lee, D. H., Luo, L., Liu, C., Kuei, C., Velumian, A. A., Butler, M. P., Brown, S. M., and Dubin, A. E. (2003) Neuronal hyperpolarization-activated pacemaker channels drive neuropathic pain, *J Neurosci 23*, 1169–1178.
47. Cao, Y. Q. (2006) Voltage-gated calcium channels and pain, *Pain 126*, 5–9.
48. Rogers, M., Tang, L., Madge, D. J., and Stevens, E. B. (2006) The role of sodium channels in neuropathic pain, *Semin Cell Dev Biol 17*, 571–581.
49. Tanaka, M., Cummins, T. R., Ishikawa, K., Dib-Hajj, S. D., Black, J. A., and Waxman, S. G. (1998) SNS Na+ channel expression increases in dorsal root ganglion neurons in the carrageenan inflammatory pain model, *Neuroreport 9*, 967–972.
50. Black, J. A., Liu, S., Tanaka, M., Cummins, T. R., and Waxman, S. G. (2004) Changes in the expression of tetrodotoxin-sensitive sodium channels within dorsal root ganglia neurons in inflammatory pain, *Pain 108*, 237–247.
51. Beyak, M. J., Ramji, N., Krol, K. M., Kawaja, M. D., and Vanner, S. J. (2004) Two TTX-resistant Na+ currents in mouse colonic dorsal root ganglia neurons and their role in colitis-induced hyperexcitability, *Am J Physiol Gastrointest Liver Physiol 287*, G845–G855.
52. Stewart, T., Beyak, M. J., and Vanner, S. (2003) Ileitis modulates potassium and sodium currents in guinea pig dorsal root ganglia sensory neurons, *J Physiol 552*, 797–807.
53. Bielefeldt, K., Ozaki, N., and Gebhart, G. F. (2002) Mild gastritis alters voltage-sensitive sodium currents in gastric sensory neurons in rats, *Gastroenterology 122*, 752–761.
54. Molliver, D. C. and Snider, W. D. (1997) Nerve growth factor receptor TrkA is down-regulated during postnatal development by a subset of dorsal root ganglion neurons, *J Comp Neurol 381*, 428–438.

55. Ogata, N. and Ohishi, Y. (2002) Molecular diversity of structure and function of the voltage-gated Na+ channels, *Jpn J Pharmacol 88*, 365–377.
56. Wu, Z. Z. and Pan, H. L. (2004) Tetrodotoxin-sensitive and -resistant Na+ channel currents in subsets of small sensory neurons of rats, *Brain Res 1029*, 251–258.
57. Amaya, F., Decosterd, I., Samad, T. A., Plumpton, C., Tate, S., Mannion, R. J., Costigan, M., and Woolf, C. J. (2000) Diversity of expression of the sensory neuron-specific TTX-resistant voltage-gated sodium ion channels SNS and SNS2, *Mol Cell Neurosci 15*, 331–342.
58. Takeda, M., Tanimoto, T., Ikeda, M., Nasu, M., Kadoi, J., Yoshida, S., and Matsumoto, S. (2006) Enhanced excitability of rat trigeminal root ganglion neurons via decrease in A-type potassium currents following temporomandibular joint inflammation, *Neuroscience 138*, 621–630.
59. Dang, K., Bielefeldt, K., and Gebhart, G. F. (2004) Gastric ulcers reduce A-type potassium currents in rat gastric sensory ganglion neurons, *Am J Physiol Gastrointest Liver Physiol 286*, G573–G579.
60. Yoshimura, N. and de Groat, W. C. (1999) Increased excitability of afferent neurons innervating rat urinary bladder after chronic bladder inflammation, *J Neurosci 19*, 4644–4653.
61. Yang, E. K., Takimoto, K., Hayashi, Y., de Groat, W. C., and Yoshimura, N. (2004) Altered expression of potassium channel subunit mRNA and alpha-dendrotoxin sensitivity of potassium currents in rat dorsal root ganglion neurons after axotomy, *Neuroscience 123*, 867–874.
62. Harriott, A. M., Dessem, D., and Gold, M. S. (2006) Inflammation increases the excitability of masseter muscle afferents, *Neuroscience 141*, 433–442.
63. Xu, G. Y., Winston, J. H., Shenoy, M., Yin, H., and Pasricha, P. J. (2006) Enhanced excitability and suppression of A-type K+ current of pancreas-specific afferent neurons in a rat model of chronic pancreatitis, *Am J Physiol Gastrointest Liver Physiol 291*, G424–G431.
64. Marder, E. and Prinz, A. A. (2002) Modeling stability in neuron and network function: the role of activity in homeostasis, *Bioessays 24*, 1145–1154.
65. Liu, L., Yang, T. M., Liedtke, W., and Simon, S. A. (2006) Chronic IL-1beta signaling potentiates voltage-dependent sodium currents in trigeminal nociceptive neurons, *J Neurophysiol 95*, 1478–1490.
66. Ramer, M. S., Thompson, S. W., and McMahon, S. B. (1999) Causes and consequences of sympathetic basket formation in dorsal root ganglia, *Pain Suppl 6*, S111–S120.
67. Ramer, M. S., Murphy, P. G., Richardson, P. M., and Bisby, M. A. (1998) Spinal nerve lesion-induced mechanoallodynia and adrenergic sprouting in sensory ganglia are attenuated in interleukin-6 knockout mice, *Pain 78*, 115–121.
68. Zhou, X. F., Deng, Y. S., Chie, E., Xue, Q., Zhong, J. H., McLachlan, E. M., Rush, R. A., and Xian, C. J. (1999) Satellite-cell-derived nerve growth factor and neurotrophin-3 are involved in noradrenergic sprouting in the dorsal root ganglia following peripheral nerve injury in the rat, *Eur J Neurosci 11*, 1711–1722.
69. Nakamura, M. and Ferreira, S. H. (1987) A peripheral sympathetic component in inflammatory hyperalgesia, *Eur J Pharmacol 135*, 145–153.
70. Chacur, M., Milligan, E. D., Gazda, L. S., Armstrong, C., Wang, H., Tracey, K. J., Maier, S. F., and Watkins, L. R. (2001) A new model of sciatic inflammatory neuritis (SIN): induction of unilateral and bilateral mechanical allodynia following acute unilateral peri-sciatic immune activation in rats, *Pain 94*, 231–244.
71. Milligan, E. D., Twining, C., Chacur, M., Biedenkapp, J., O'Connor, K., Poole, S., Tracey, K., Martin, D., Maier, S. F., and Watkins, L. R. (2003) Spinal glia and proinflammatory cytokines mediate mirror-image neuropathic pain in rats, *J Neurosci 23*, 1026–1040.
72. Dobretsov, M., Hastings, S. L., Stimers, J. R., and Zhang, J.-M. (2001) Mechanical hyperalgesia in rats with chronic perfusion of lumbar dorsal root ganglion with hyperglycemic solution, *J Neurosci Methods 110*, 9–15.

Chapter 7

Animal Models of Central Neuropathic Pain

Bryan Hains and Louis P. Vera-Portocarrero

Abstract

Central neuropathic pain is triggered by trauma, neurological disease, or infection of the central nervous system. A number of powerful models of central pain produce consistent phenotypes of behavioral hypersensitivity and provide an understanding into the mechanisms underlying persistent changes in nociceptive processing. Here we discuss behavioral phenomenology, how changes in neuronal firing properties lead to system-wide modulation of information processing, and highlight relevant animal models associated with spinal cord injury, multiple sclerosis, and infection – key contributors to central pain.

1. Introduction: Centrally Generated Pain

Central neuropathic pain arises from injury to the central nervous system (CNS) as the consequence of direct trauma to the spinal cord or brain, or a secondary response to disease or illness. In either case the common feature is the emergence of changes in how the CNS processes nociceptive information at the molecular, cellular, and/or circuit level, leading to system-wide changes in neuronal excitability and behavioral manifestations such as allodynia or hyperalgesia.

In this light, centrally located mechanisms account for the altered behaviors. There are several that could provide the conditions for the observed changes, including the unmasking of previously ineffective latent pathways, decreased inhibition, the development of central sensitization or denervation hypersensitivity, and/or changes in the excitatory amino acid, peptide or other receptor population. Additionally, structural alterations may take place over days such as sprouting of the central processes of primary afferents, and plastic changes in molecular and/or neuroanatomical pathways induced by the resultant denervation.

Chao Ma and Jun-Ming Zhang (eds.), *Animal Models of Pain*, Neuromethods, vol. 49,
DOI 10.1007/978-1-60761-880-5_7, © Springer Science+Business Media, LLC 2011

Alterations in the expression of neurotransmitter systems, ion channels, and activation of glial cells can also contribute to the development and maintenance of neuronal hyperexcitability leading to central pain.

2. Nociceptive Withdrawal: Hyperreflexia

A discussion of central pain would be incomplete without mention of a point of contention among behavioralists: reflex changes after central injury. In pain research all measurements are indicators of nociception in response to a standardized stimulus but never a pain perception as humans might experience it. Sherrington first observed and characterized the ipsilateral flexion and extension in the contralateral limb as a mechanism to preserve balance, and associated this system with flight responses in the unaffected limbs to escape potential tissue damage. The thresholds of tactile and nociceptive input and pain perception are related, and there is a strong correlation between pain intensity and reflex response in humans that permits the use of this reflex as a quasi-objective measure of experimental pain (64). At the systems level, the nociceptive withdrawal reflex is used as a physiological measure of pain signaling.

There is a concern, however, that behavioral measurement of an injury to the CNS is merely reflexive in nature. That is, because descending control systems may be damaged, paw flexion responses are not a true indicator of pain but one of heightened spinal reflexes. This may be true to some degree, but evidence showing enhanced electrical excitability of dorsal horn and thalamic nociceptive neurons, as well as changes in cortical structures involved in interpreting nociceptive signals, indicate that nociceptive signals do indeed reach supraspinal areas involved in pain processing. In addition, biting and head turning in association with vocalizations and other complex supraspinal behaviors in response to peripheral stimulation support the notion that CNS injuries result in more than simple hyperreflexia.

3. Central Sensitization and Hyperexcitability

Central sensitization is perhaps the most common result of CNS injury or disease, resulting in facilitation of excitatory synaptic responses and depressed inhibition, thereby amplifying responses to innocuous and noxious inputs (65). These changes may be restricted to the activated synapse, but can also spread to adjacent synapses (an underlying cause of secondary hyperalgesia). Increased gain results in recruitment of these inputs, causing them to fire to normally ineffective or below-threshold inputs (54). These changes are responsible for pain produced by low-threshold afferent inputs.

In the particular case of multireceptive or wide dynamic range dorsal horn neurons, changes in central sensitization correspond with changes that occur in chronic pain. These alterations result in the development of sustained hyperexcitability of central nociceptive neurons within the spinal dorsal horn – particularly of wide dynamic range dorsal horn neurons (WDR), many of which are nociceptive projection neurons that comprise the spinothalamic tract (STT) (32). Extracellular recordings from the spinal dorsal horn reveals shifts in proportions of cells responding to noxious peripheral stimulation, alterations (increases and irregularity) in spontaneous background activity, increased evoked activity to (formerly) innocuous and noxious stimuli, and increases in afterdischarge activity following stimulation. Similar observations have been made in the thalamus (21, 22). This parallelism has been described in both rodent and primate models of central pain.

4. Spinal Cord Injury

Direct injury to the spinal cord (SCI) represents perhaps the most frequently utilized model of central neuropathic pain, and at least four distinct models have been developed over the last two decades. In this section the contusion, hemisection, excitotoxic, and ischemic models of SCI are described.

4.1. Contusion

Spinal contusion lesions are perhaps the oldest and most widely used method of producing central neuropathic pain. Contusion SCI may best parallel the injury profile described in human SCI (7). The development and widespread use of the MASCIS impactor developed by Wise Young (W. M. Keck Center for Collaborative Neuroscience in Piscataway, New Jersey; http://www.keck.rutgers.edu/MASCIS/mascis.html) has helped to standardize the contusion model of SCI (4, 10, 19), and to characterize the acute, secondary, and chronic injury processes that result from injury. Since these initial descriptions, several recent methods have been developed to control with a greater degree of precision the contusion process itself.

Contusion SCI is typically produced at spinal segment T9 to avoid direct damage to central pattern generator circuitry. A laminectomy spanning a single vertebral segment is created and the MASCIS device is used by dropping a 2.0-mm-diameter rod (10 gm) from a preset height of up to 50 mm above the surface of the spinal cord (25 mm is most commonly used). Impact parameters (impact velocity, cord compression, and cord displacement) are measured by potentiometers attached to the vertebral column to ensure repeatability of the impact. After surgery, the overlying muscle and skin is closed. With contusion SCI, since micturition is impaired, bladders require manual expression twice daily until reflex bladder function returns, usually within 10 days after injury.

A newer injury device, referred to as the Infinite Horizons (IH) impactor (Precision Systems and Instrumentation, LLC, Lexington, KY; http://www.presysin.com), has recently been developed (52). The advantage provided by the IH device is the ability to produce a force-defined SCI. An additional benefit of the IH device is that it eliminates the "bounce" associated with free weight drops produced by the MASCIS device.

With both devices, moderate spinal contusion results in an incomplete SCI that is consistent in lesion volume. Furthermore, spinal contusion produces permanent chronic central pain syndromes (20, 22, 23, 26, 43, 70) in as many as 87% of all injured animals (43). Animals that exhibit both mechanical allodynia and thermal hyperalgesia also demonstrate a bilateral band of allodynia at, and rostral to, the segmental level of injury (26). This observation is in agreement with other models of SCI and of patient reports.

4.2. Hemisection

Claire Hulsebosch first developed and characterized the unilateral hemisection model of chronic pain following lateral spinal hemisection, an equivalent of a Brown-Sequard syndrome (8, 9). In this model the left side of the spinal cord is unilaterally hemisected at T13 by the following procedure: following palpation of the dorsal surface to locate the rostral borders of the sacrum and dorsal spinous processes of the lower thoracic and lumbar vertebrae, the T11–T12 laminae are determined by locating the last rib, which attaches to the rostral end of the T13 vertebra. A laminectomy is performed at the T11 vertebra, the lumbar spinal cord is identified with accompanying dorsal vessel, and the spinal cord is hemisected at spinal segment T13 with a No. 11 scalpel blade without damage to the posterior spinal blood vessel or branches. Alternatively, spring loaded ophthalmic microscissors can be used to cut the cord. Muscle and fascia are sutured and skin closed. The extent of the hemisection lesion, assessed histologically, should be confined unilaterally and include the dorsal column, Lissauer's tract, lateral and ventral column systems, and gray matter.

The hemisection model has the advantages of consistency in response as the highest number of animals develop the same amount of allodynia (43). This model has several additional benefits: surgical ease (no expensive and sophisticated weight drop apparatus is needed), surgical selectivity that avoids the variability of different lesions, ease of reproducibility, avoidance of profound hind limb locomotor deficits that occur after contusion, and elimination of the twice daily bladder expressions associated with SCI with accompanying bladder infections.

In the hemisection model, nociceptive transmission is preserved through remaining and/or modified spinal cord circuitry; however, in regard to the hemisection model, it might be argued that the finding of an enhanced paw withdrawal response is not useful as an indicator of nociceptive behavior with concomitant supraspinal perception – reasoning based on the fact that hemisection interrupts some pathways involved in supraspinal transmission of somatosensory information. It must be remembered, however, that this reflex occurs independently of ascending or descending pathways since complete transections or decerebration results in heightened flexor reflex activity. This would lead to the development of behavioral changes which merely reflect heightened reflexes of the classical upper motor neuron lesion, partially due to the loss of descending inhibition leading to enhanced segmental spinal excitability. While this can account for the acute and chronic effects of increased withdrawal of the hind limb ipsilateral to the lesion, it does not explain the attenuated mechanical and radiant heat thresholds for limb withdrawal to peripheral stimuli on the contralateral hind limb and bilaterally in the forelimbs. The added behavior of licking and attending to each hind limb, head turns, and vocalizations following stimulus application addresses the involvement of supraspinal transmission of the same primary afferent input which elicited the withdrawal reflex. Many pathways exist that can permit the ascent of noxious somatosensory information outside of conventionally accepted pathways.

4.3. Excitotoxic

In another SCI model, focal excitotoxic lesions are created within the spinal cord using the AMPA–metabotropic receptor agonist quisqualic acid (67–69). This model is based on evidence that excitatory amino acids play a key role in the underlying neurodegenerative and proinflammatory pathology caused by contusion SCI. Quisqualic acid produces similar morphological changes such as cavitation, cyst formation, reactive gliosis, and cell death observed after contusion but without damage to en passant fiber tracts at the lesion level.

The excitotoxic model is produced by microinjecting quisqualic acid at predetermined depths below the surface of the spinal cord. After the vertebral column is exposed, the spinous process and vertebral lamina are removed from one spinal level and the dura incised longitudinally and reflected unilaterally. At sites where injections are to be made the pia matter is carefully elevated using No. 5 Dumont forceps and a small hole is created to allow penetration of the injection pipette. Injections are made in a single segment at spinal levels typically ranging from T10 to L4.

Glass micropipettes (tip diameter 5–10 μm) attached to a Hamilton microliter syringe (volume 5 μl) are used for injections. The syringe is mounted on a microinjector attached to a micromanipulator. Injections are made between the dorsal vein and dorsal root entry zone at depths ranging from 300 to 1,200 μm below the surface of the spinal cord. Stock solutions of 125 mM QUIS are made in normal saline. At each injection site 0.2 μl of QUIS is injected over a 60 s time interval. The total volume of QUIS typically ranges from 0.6 to 1.2 μl (one to two depths/track; three tracks/animal separated by 0.3 mm). Based on the volumes and concentration of QUIS injected, the total quantity of QUIS delivered in each animal ranges from 13.8 to 27.5 mg. Following injections muscles were closed in layers.

The key benefit of the excitotoxic model is that lesions can be very precisely created in very specific locations. This selective lesioning permits surgical targeting of specific pathways and/or spinal cord structures a few microns in diameter. At the same time, larger doses can produce larger scale injuries that occupy multi-millimeter areas of tissue.

In the quisqualic acid excitotoxic lesion model, hyperexcitability is evident in spinal neurons and overgrooming/autotomy behavior develops in dermatomes that border the lesion site. Ninety percent of the animals with damage in the dorsal horn (preserving the superficial laminae) develop spontaneous excessive grooming behavior and all animals with QUIS injections develop varying degrees of hypersensitivity to mechanical and thermal stimuli.

4.4. Ischemia

Spinal ischemia is less frequently used to create a lesion in the spinal cord that results in central pain (24, 66). Spinal ischemia is initiated by vascular occlusion resulting from the interaction between the photosensitizing dye Erythrosin B and an argon laser beam. The laser is directed at the spinal cord dorsal horn at a time when a solution containing a photopigment is introduced into the circulation. An intravascular photochemical reaction occludes blood vessels, thereby producing subsequent spinal cord ischemia.

To produce the photochemical insult, Erythrosin B (32.5 mg/kg, Red No. 3, Aldrich–Chemic, Steinheim, Germany) is dissolved in 0.9% saline and injected i.v. Immediately following the injection, rats are irradiated for 1 min under the laser beam at vertebral segment T10 (spinal segment T11–T12). A tunable argon ion laser (Innova, Model 70, Coherent Laser Products Division, Palo Alto, CA) operating at 514.5 nm (near the absorption maximum of Erythrosin B) is used for irradiation. The vertebra and the spinal cord are irradiated with an average power of 0.16 W delivered via a beam chopper operating at 500 Hz (with 2.4 W peak power, 6.7% duty cycle). A knife edge beam is used to cover the single vertebra. To reduce the amount of heat introduced into the tissues during the process of irradiation, two piezoelectric fans (Piezo Systems, Inc., Cambridge, MA) operating at peak air velocity of 6 m/s are placed as close to the spinal column as possible (approximately 5 mm away).

In a behavioral characterization 6 days after surgery (24), ~75% of injured rats exhibited strong allodynia in response to innocuous mechanical stimulation of the caudal trunk, hind limbs, and hind paws, which were areas innervated by the ischemic spinal segments. Hypersensitivity to cold stimuli was also observed. No changes in sensitivity to noxious heat were detected with the hot-plate test.

5. Multiple Sclerosis

Multiple sclerosis (MS) is an autoimmune disease that affects 2.5 million people worldwide. The main characteristic of this disease is myelin loss and axonal degeneration. Some of the main symptoms of MS are trunk and limb paresthesias, limb weakness, and gait ataxia (45). Pain is another common characteristic of MS. Studies have determined that from 30 to 90% of patients with MS experience pain during the course of the disease (2). In about 8% of patients, pain is the first presenting symptom of the disease (17). The pain of MS can be debilitating (18) and it is characterized by cold allodynia and mechanical hyperalgesia (56). Until a few years ago, there was a lack of attention to pain symptoms in preclinical models of MS. Recently two animal models of MS have been validated as models which present some of the pain characteristics of the disease.

Experimental Autoimmune Encephalomyelitis (EAE): EAE is a frequently used animal model of MS (46, 47). The disease can be induced in a variety of rodent species and strains (28). There are several types of EAE induced in different mice strains. The ones used to study nociceptive responses are as follows.

5.1. EAE in the SJL Strain of Mice

In this strain of mice, EAE is induced by subcutaneous injection of 200 μl of a solution containing 150 μg of myelin proteolipid protein peptide 139–151 ($PLP_{139-151}$). The goal of this inoculation is to produce an immune response to PLP, which results in chronic demyelination (6, 38). This method is known as the "active" method since it requires for the organism to produce antibodies against the injected PLP. On the other hand, the "passive" method requires the injection of activated splenocytes, previously activated with PLP, to induce EAE. Both methods have been proven to produce similar symptoms of demyelination (5, 6).

The first study to study nociceptive responses in this disease was in 1986 (48). In this study, actually done in the Lewis strain of rats, nociceptive responses of the tail were investigated. It was found that tail sensory thresholds were impaired, giving rise to longer tail flick latencies. This was to be expected since the same study found conduction block of Aδ fibers in the sacral DRG induced by demyelination. Nonetheless, the chronic course of nociceptive responses would not be studied until later.

In a study using the EAE model in mice, investigators characterized nociceptive responses during the chronic phase of the disease (1). The study demonstrated that after the initial increase in withdrawal latency of the tail (hypoalgesia), the latency decreased (hyperalgesia), significantly up to 38 days after immunization (1). Most importantly, there was a direct correlation between the site of demyelination and where the hyperalgesia was observed. Importantly, the onset of hyperalgesia preceded the onset of motor deficits seen in this murine model; therefore, the change in nociceptive responses cannot entirely be explained by the motor impairments. Therefore, a change in sensory thresholds is more likely.

The mechanisms behind increased nociceptive responses are still largely unknown. Few studies have addressed this issue but the scant evidence points to the damage of different nerve pathways as a source of increased excitability and subsequent hyperalgesia. A loss of descending noradrenergic fibers has been noted in mice with EAE (63). These fibers are involved in descending inhibition of nociceptive transmission at the spinal cord level (3, 16). Loss of these fibers would give rise to increased dorsal horn excitability and subsequent hyperalgesia (39).

Another intriguing possibility is the possible role of glial cells in nociception. The loss of myelin induced by the immunization process gives rise to glial cells activation (46). Glial cells are thought to be drivers of nociceptive sensitization (61, 62), therefore glial cells might be important players in the initiation and maintenance of nociceptive hypersensitivity in animal models of EAE and possibly MS in humans.

5.2. Thelier's Murine Encephalomyelitis Virus in the SJL Strain of Mice

This experimental model of MS is obtained by intracerebral inoculation of Daniels (DA) strain of Thelier's Murine Encephalomyelitis Virus (TMEV) (51). This treatment produces a biphasic disease. An early phase last from 3 to 12 days after injection, followed by a late chronic demyelinating disease, which develops between 30 and 40 days postinjection (13). The early phase is characterized by infiltration of T lymphocytes (44). In the late phase, mice injected with TMEV develop chronic demyelinating disease with extensive lesions of the white matter and mononuclear infiltrates in the spinal cord (51). This chronic phase leads to progressive spinal cord atrophy and axonal loss accompanied by disruption in motor coordination, hind limb paralysis, spasticity, ataxia, and incontinence (14, 37).

Nociceptive responses are altered in the TMEV model. Mice infected with TMEV present thermal hyperalgesia and mechanical allodynia (34). The alterations in sensory thresholds preceded the onset of motor deficits, once again ruling out the possibility that these mice have impaired capability to respond to noxious stimulation. This study identified increased density of PGP9.5-ir and

CGRP-ir fibers in TMEV mice when compared to controls (34). These results suggest that increased sprouting of nerve fibers in the periphery might be involved in the mechanisms driving increased nociceptive hypersensitivity in this murine model of MS.

6. CNS Infection

Infections of the CNS can produce pain. Many of the infections that attack the nervous system have both a central and peripheral components. The most common infections that affect the CNS and which are accompanied by pain symptoms are postherpetic neuralgia (27), HIV infection (58) and meningitis (49). The first two types of infections affect peripheral and CNS, so they cannot be classified as central neuropathic pain syndromes exclusively. Meningitis comprises infection and inflammation of the meninges surrounding the CNS, one of the consequences of this process being pain. Since meningitis affects, in most cases, the meninges surrounding the brain, this leads to headache pain, which can then be considered central neuropathic pain. There are few models of infection-driven central neuropathic pain. They mostly comprised of injection of infecting agents into the CNS to produce headaches (when injected in cranial structures) or in the spinal cord to produce behavioral hypersensitivity.

6.1. Central Application of Infecting Agents as an Animal Model of Migraine

It is estimated that up to 30 million people in the USA suffer from migraine headaches (33). The pathophysiology of migraine is not fully understood. Many theories have been put forward to try to explain the mechanisms behind the production of migraine headache including vascular mechanisms (15), central mechanisms (12) and medication overuse (11). Based upon the neurovascular model, injection of inflammatory agents has been used to mimic some of the aspects of migraine in animals. One model injects the endotoxin lipopolysaccharide (LPS) into the intracerebroventricular space (ICV injection). Application of LPS in the brain produces an inflammatory response characterized by infiltration of monocytes and activation of glial cells (50). The procedure itself consists of injection into the lateral ventricle of 2.4 μg of LPS in a volume of 10 μl under isofluorane anesthesia (36). After 3.5 h, the rats were tested by application of air jets to the face of the rats and measuring ultrasound vocalization (USV). The rats were also tested with a cold stimulus to the face. Rats injected with LPS had lower thresholds to vocalization after application of the air jet to the face. Similarly, rats treated with LPS presented cold allodynia (36). These hypersensitive responses were attenuated by different analgesic drugs including morphine, diazepam, and antimigraine drugs like triptans (36).

6.2. Spinal Application of Infecting Agents as an Animal Model of HIV-Induced Central Pain

As many as 25–50% of patients with AIDS suffer of painful neuropathies (58). Many investigators have sought to develop animal models of HIV-induced neuropathies. The most commonly used model is obtained by application of gp120, an envelope glycoprotein expressed by the HIV-1 virus (55). Injection of gp120 into peripheral nerves has been used to mimic distal symmetric polyneuropathy, a subset of HIV-induced neuropathy (35, 59, 60). Intrathecal injection of gp120 into the spinal cord has been used to emulate central neuropathic pain resulting from HIV infection.

Intrathecal injection of gp120 causes mechanical and thermal hypersensitivity (40) and promotes the release of proinflammatory cytokines in the spinal cord (41). The main mechanism by which gp120 produces mechanical allodynia and thermal hyperalgesia is related to its activating properties of glial cells and subsequent activation of inflammatory cascades. Intrathecal application of gp120 activates astrocytes and microglia via chemokine receptors (29). Activation of astrocytes and microglia is necessary for the pain-facilitating action of gp120 (30, 40). Spinal application of gp120 activates the p38 MAP kinase pathways to facilitate pain (42) and it can also activate the nitric oxide pathway (25). Downstream from glia activation, gp120 activates the complement cascade and the NFkB pathway (31, 57). The overall picture emerging is that gp120 can activate inflammatory cascades which gives rise to a state of pain maintained by proinflammatory cytokines (53).

7. Summary

Central neuropathic pain is defined by alterations in processing of the CNS. The injury can be induced by trauma, neurological disease, or infection. Pain caused by SCI, by far, has been the most studied entity and from which most of our knowledge of central pain comes from. There are a great variety of SCI models, ranging from contusion to hemisection of the spinal cord to excitotoxic and ischemic insults to the cord. All these models produce consistent phenotypes of behavioral hypersensitivity and they give powerful tools to dissect the mechanisms driving the maintenance of chronic pain after SCI. Models of neurological disease are not usually used as models of pain. One exception is the use of models of MS to study nociceptive responses. This investigative drive is still in its infancy as there are few studies addressing pain phenotypes in MS models. Finally, the study of CNS infection has given birth to models of migraine and HIV-induced central neuropathies, which has provided a great deal of information regarding the mechanisms which drive pain facilitation after an infecting agent is introduced in the CNS. At the other end of the spectrum, there is a need to use animal models of stroke and Parkinson's disease to study pain phenotypes and be able to help in the future patient populations that have pain after a central stroke or during Parkinson's disease.

References

1. Aicher SA, Silverman MB, Winkler CW, Bebo Jr BF. Hyperalgesia in an animal model of multiple sclerosis. Pain 2004; 110:560–570.
2. Archibald CJ, McGrath PJ, Ritvo PG, Fisk JD, Bhan V, Maxner CE, et al. Pain prevalence, severity and impact in a clinical sample of multiple sclerosis patients. Pain 1994; 58:89–93.
3. Basbaum AI, Clanton CH, Fields HL. Three bulbospinal pathways from the rostral medulla of the cat: an autoradiographic study of pain modulating systems. J Comp Neurol 1978; 178:209–224.
4. Basso DM, Beattie MS, Bresnahan JC. Graded histological and locomotor outcomes after spinal cord contusion using the NYU weight-drop device versus transection. Exp Neurol 1996; 139(2):244–256.
5. Bebo Jr BF, Vandenbark AA, Offner H. Male SJL mice do not relapse after induction of EAE with PLP 139–152. J Neurosci Res 1996; 45:680–689.
6. Bourdette DN, Vandenbark AA, Meshul C, Whitham R, Offner H. Basic protein-specific T-cell lines that induce experimental autoimmune encephalomyelitis in SJL/J mice: comparison with Lewis rat lines. Cell Immunol 1988; 112:351–363.
7. Bunge RP, Puckett WR, Becerra JL, Marcillo A, Quencer RM. Observations on the pathology of human spinal cord injury. A review and classification of 22 new cases with details from a case of chronic cord compression with extensive focal demyelination. Adv Neurol 1993; 59:75–89.
8. Christensen MD, Everhart AW, Pickelman JT, Hulsebosch CE. Mechanical and thermal allodynia in chronic central pain following spinal cord injury. Pain 1996; 68(1):97–107.
9. Christensen MD, Hulsebosch CE. Chronic central pain after spinal cord injury. J Neurotrauma 1997; 14(8):517–537.
10. Constantini S, Young W. The effects of methylprednisolone and the ganglioside GM1 on acute spinal cord injury in rats. J Neurosurg 1994; 80(1):97–111.
11. Cupini LM, Calabresi P. Medication-oversue headache: pathophysiological insights. J Headache Pain 2005; 6(4):199–202.
12. Cutrer FM. Pathophysiology of migraine. Semin Neurol 2006; 26(2):171–180.
13. Del Canto MC, Kim BS, Miller SD, Melvold RW. Theiler's murine encephalomyelitis virus (TMEV)-induced demyelination: a model for human multiple sclerosis. Methods 1996; 10:453–461.
14. Drescher KM, Pease LR, Rodriguez M. Antiviral immune responses modulate the nature of central nervous system (CNS) disease in a murine model of multiple sclerosis. Immunol Rev 1997; 159:177–193.
15. Feniuk W, Humphrey PP, Perren MJ, Connor HE, Whalley ET. Rationale for the use of 5-HT1-like agonists in the treatment of migraine. J Neurol 1991; 238(Suppl 1):s57–s61.
16. Fields HL, Heinricher MM, Mason P. Neurotransmitters in nociceptive modulatory circuits. Annu Rev Neurosci 1991; 14: 219–245.
17. Fryze W, Zaborski J, Czlonkowska A. Pain in the course of multiple sclerosis. Neurol Neurochir Pol 2002; 36:275–284.
18. Ehde DM, Gibbons LE, Chwastiak L, Bonbardier CH, Sullivan MD, Kraft GH. Chronic pain in a large community sample of persons with multiple sclerosis. Mult Scler 2003; 9:605–611.
19. Gruner JA. A monitored contusion model of spinal cord injury in the rat. J Neurotrauma 1992; 9(2):123–126.
20. Hains BC, Klein JP, Saab CY, Craner MJ, Black JA, Waxman SG. Upregulation of sodium channel Nav1.3 and functional involvement in neuronal hyperexcitability associated with central neuropathic pain after spinal cord injury. J Neurosci 2003; 23(26): 8881–8892.
21. Hains BC, Saab CY, Waxman SG. Alterations in burst firing of thalamic VPL neurons and reversal by Na(v)1.3 antisense after spinal cord injury. J Neurophysiol 2006; 95(6): 3343–3352.
22. Hains BC, Saab CY, Waxman SG. Changes in electrophysiological properties and sodium channel Nav1.3 expression in thalamic neurons after spinal cord injury. Brain 2005; 128(Pt 10):2359–2371.

23. Hains BC, Waxman SG. Activated microglia contribute to the maintenance of chronic pain after spinal cord injury. J Neurosci 2006; 26(16):4308–4317.
24. Hao JX, Xu XJ, Aldskogius H, Seiger A, Wiesenfeld-Hallin Z. Allodynia-like effects in rat after ischaemic spinal cord injury photochemically induced by laser irradiation. Pain 1991; 45(2):175–185.
25. Holguin A, O'Connor KA, Biedenkapp J, Campisi J, Wieseler-Frank J, Milligan ED, Hansen MK, Spataro L, Maksimova E, Bravmann C, Martin D, Fleshner M, Maier SF, Watkins LR. HIV-1 gp120 stimulates proinflammatory cytokine-mediated pain facilitation via activation of nitric oxide synthase-I (nNOS). Pain 2004; 110(3):517–530.
26. Hulsebosch CE, Xu GY, Perez-Polo JR, Westlund KN, Taylor CP, McAdoo DJ. Rodent model of chronic central pain after spinal cord contusion injury and effects of gabapentin. J Neurotrauma 2000; 17(12):1205–1217.
27. Jensen TS, Finnerup NB. Management of neuropathic pain. Curr Opin Support Palliat Care 2007; 1(2):126–131.
28. Joy J, Johnston RJ. Multiple sclerosis: current status and strategies for the future. Washington, DC: National Academy Press; 2000.
29. Koka P, He K, Zack JA, Kitchen S, Peacock W, Fried I, Tran T, Yashar SS, Merrill JE. Human immunodeficiency virus-1 envelope proteins induce interleukin 1, tumor necrosis factor alpha, and nitric oxide in glial cultures derived from fetal, neonatal and adult human brain. J Exp Med 1995; 182:941–951.
30. Ledeboer A, Sloane EM, Milligan ED, Frank MG, Mahony JH, Maier SF, Watkins LR. Minocycline attenuates mechanical allodynia and proinflammatory cytokine expression in rat models of pain facilitation. Pain 2005; 115(1–2):71–83.
31. Ledeboer A, Gamanos M, Lai W, Martin D, Maier SF, Watkins LR, Quan N. Involvement of spinal cord nuclear factor kappaB activation in rat models if proinflammatory cytokine-mediated pain facilitation. Eur J Neurol 2005; 22(8):1977–1986.
32. Lin Q, Peng YB, Willis WD. Inhibition of primate spinothalamic tract neurons by spinal glycine and GABA is reduced during central sensitization. J Neurophysiol 1996; 76(2):1005–1014.
33. Lipton RB, Stewart WF, Scher AI. Epidemiology and economic impact of migraine. Curr Med Res Opin 2001; 17(Suppl 1):s4–s12.
34. Lynch JL, Gallus NJ, Ericson ME, Beitz AJ. Analysis of nociception, sex and peripheral nerve innervation in the TMEV animal model of multiple sclerosis. Pain 2008; 136:293–304.
35. Maratou K, Wallace VC Hasnie FS, Okuse K, Hosseini R, Jina N, Blackbeard J, Pheby T, Orengo C, Dickenson AH, McMahon SB, Rice AS. Comparison of dorsal root ganglion gene expression in rat models of traumatic HIV-associated neuropathic pain. Eur J Pain 2009; 13(4):387–398.
36. Martino G, Perkins MN. Tactile-induced ultrasonic vocalization in the rat: a novel assay to assess anti-migraine therapies in vivo. Cephalalgia 2008; 28(7):723–733.
37. McGavern DB, Murray PD, Rivera-Quinones C, Schmelzer JD, Low PA, Rodriguez M. Axonal loss results in spinal cord atrophy, electrophysiological abnormalities and neurological deficits following demyelination in a chronic inflammatory model of multiple sclerosis. Brain 2000; 123:519–531.
38. McRae BL, Kennedy MK, Tan LJ, Dal Canto MC, Picha KS, Miller SD. Induction of active and adoptive relapsing experimental autoimmune encephalomyelitis (EAE) using an encephalitogenic epitope of proteolipid protein. J Neuroimmunol 1992; 38:229–240.
39. Millan MJ. Descending control of pain. Prog Neurobiol 2002; 66(6):355–474.
40. Milligan ED, Mehmert KK, Hinde JL, Harvey LO, martin D, Tracey KJ, Maier SF, Watkins LR. Thermal hyperalgesia and mechanical allodynia produced by intrathecal administration of the human immunodeficiency virus-1 (HIV-1) envelope glycoprotein, gp120. Brain Res 2000; 861(1):105–116.
41. Milligan ED, O'Connor KA, Nguyen KT, Armstrong CB, Twining C, Gaykema RP, Holguin A, Martin D, Maier SF, Watkins LR. Intrathecal HIV-1 envelope glycoprotein gp120 induces enhanced pain states mediated by spinal cord proinflammatory cytokines. J Neurosci 2001; 21(8):2808–2819.
42. Milligan ED, O'Connor KA, Armstrong CB, Hansen MK, Martin D, Tracey KJ, Maier SF, Watkins LR. Systemic administration of CNI-1493, a p38 mitogen-activated protein kinase inhibitor, blocks intrathecal human immunodeficiency virus-1 gp120-induced enhanced pain states in rats. J Pain 2001; 2(6):326–333.
43. Mills CD, Hains BC, Johnson KM, Hulsebosch CE. Strain and model differences in behavioral outcomes after spinal cord injury in rat. J Neurotrauma 2001; 18(8):743–756.
44. Oleszak EL, Chang JR, Friedman H, Katsetos CD, Platsoucas CD. Thelier's virus infection: a model of multiple sclerosis. Clin Microbiol Rev 2004; 17(1):174–207.

45. Osterberg A, Boivie J, Thoumas K-A. Central pain in multiple sclerosis – prevalence and clinical characteristics. Eur J Pain 2005; 9:531–542.
46. Paterson PY. Experimental allergic encephalomyelitis and autoimmune disease. Adv Immunol 1966; 5:131–208.
47. Paterson PY, Swanborg RH. Demyelinating diseases of th central and peripheral nervous system. In: Sampter M, Talmage DW, Frank MM, Austen KF, Claman HN, editors. Immunological disease. Boston: Little Brown and Col.; 1988. pp. 1877–1916.
48. Pender MP. Ascending impairment of nociception in rats with experimental allergic encephalomyelitis. J Neurol Sci 1986; 75:317–328.
49. Perkins AT, Ondo W. When to worry about headache. Head pain as a clue to intracranial disease. Postgrad Med 1995; 98(2):197–201.
50. Perry VH, Andersson PB. The inflammatory response in the CNS. Neuropathol Appl Neurobiol 1992; 18(5):454–459.
51. Rodriguez M, Leibowitz JL, Lampert PW. Persistent infection of oligodendrocytes in Theiler's virus-induced encephalomyelitis. Ann Neurol 1993; 3:426–433.
52. Scheff SW, Rabchevsky AG, Fugaccia I, Main JA, Lumpp JE Jr. Experimental modeling of spinal cord injury: characterization of a force-defined injury device. J Neurotrauma 2003; 20(2):179–193.
53. Schoeniger-Skinner DK, Ledeboer A, Frank MG, Milligan ED, Poole S, Martin D, Maier SF, Watkins LR. Interleukin-6 mediates low-threshold mechanical allodynia induced by intrathecal HIV-1 envelope glycoprotein gp120. Brain Behav Immun 2007; 21(5):660–667.
54. Simone DA, Sorkin LS, Oh U, Chung JM, Owens C, LaMotte RH, Willis WD. Neurogenic hyperalgesia: central neural correlates in responses of spinothalamic tract neurons. J Neurophysiol 1991; 66(1):228–246.
55. Sundar SK, Cierpial MA, Kamaraju LS, Long S, Hsieh S, Lorenz C, Aaron M. Ritchie JC, Weiss JM. Human immunodeficiency virus glycoprotein (gp120) infused into rat brain induces interleukin v1 to elevate pituitary–adrenal activity and decrease peripheral cellular immune responses. Proc Natl Acad Sci U S A 1991; 88:11246–11250.
56. Svendsen KB, Jensen TS, Overvad K, Hansen HJ, Koch-Henriksen N, Bach FW. Pain in patients with multiple sclerosis: a population-based study. Arch Neurol 2003; 60:1089–1094.
57. Twining CM, Sloane EM, Schoeniger DK, Milligan ED, martin D, Marsch H, Maier SF, Watkins LR. Activation of the spinal cord complement cascade might contribute to mechanical allodynia induced by three animal models of spinal sensitization. J Pain 2005; 6(3):174–183.
58. Verma S, Estanislao L, Simpson D. HIV-associated neuropathic pain: epidemiology, pathophysiology and management. CNS Drugs 2005; 19(4):325–334.
59. Wallace VC, Blackbeard J, Segerdahl AR, Hasnie F, Pheby T, McMahon SB, Rice AS. Characterization of rodent models of HIV-gp120 and anti-retroviral-associated neuropathic pain. Brain 2007; 130:2688–2702.
60. Wallace VC, Blackbeard J, Pheby T, Segerdahl AR, Davies M, Hasnie F, Hall S, McMahon SB, Rics AS. Pharmacological, behavioral and mechanistic analysis of HIV-1 gp120 induced painful neuropathy. Pain 2007; 133(1–3):47–63.
61. Watkins LR, Milligan ED, Maier SF. Glial activation: a driving force of pathological pain. Trends Neurosci 2001; 24:450–455.
62. Watkins LR, Maier SF. Beyond neurons: evidence that immune and glial cells contribute to pathological pain states. Physiol Rev 2002; 82:981–1011.
63. White SR, Vyas D, Bieger D, Samathanam G. Monoamine-containing fiber plexuses in the spinal cord of guinea pigs during paralysis, recovery and relapse stages of chronic relapsing experimental allergic encephalomyelitis. J Neuroimmunol 1989; 22:211–221.
64. Willer JC, De Broucker T, Le Bars D. Encoding of nociceptive thermal stimuli by diffuse noxious inhibitory controls in humans. J Neurophysiol 1989; 62(5):1028–1038.
65. Woolf CJ, Salter MW. Neuronal plasticity: increasing the gain in pain. Science 2000; 288(5472):1765.
66. Xu XJ, Hao JX, Aldskogius H, Seiger A, Wiesenfeld-Hallin Z. Chronic pain-related syndrome in rats after ischemic spinal cord lesion: a possible animal model for pain in patients with spinal cord injury. Pain 1992; 48(2):279–290.
67. Yezierski RP, Liu S, Ruenes GL, Kajander KJ, Brewer KL. Excitotoxic spinal cord injury: behavioral and morphological characteristics of a central pain model. Pain 1998; 75(1):141–155.
68. Yezierski RP, Park SH. The mechanosensitivity of spinal sensory neurons following intraspinal injections of quisqualic acid in the rat. Neurosci Lett 1993; 157(1):115–119.
69. Yezierski RP. Pain following spinal cord injury: pathophysiology and central mechanisms. Prog Brain Res 2000; 129:429–449.
70. Zhao P, Waxman SG, Hains BC. Extracellular signal-regulated kinase-regulated microglia-neuron signaling by prostaglandin E2 contributes to pain after spinal cord injury. J Neurosci 2007; 27(9):2357–2368.

Chapter 8

Animal Models of Cancer Pain

Paul W. Wacnik, Cholawat Pacharinsak, and Alvin J. Beitz

Abstract

The incidence of cancer pain is high in patients with advanced disease as well as in patients undergoing active treatment for solid tumors. Further, modern cancer therapies have significantly increased survival rates, making effective pain control critical as unrelieved pain significantly decreases the quality of life of such patients. Thus, the goal of pain management is to not only alleviate pain, but also maintain the patient's normal quality of life. To meet this challenge, novel analgesics with greater efficacy but fewer side effects are needed for alleviating cancer-induced pain. Recent advances in understanding the mechanism(s) of cancer pain have been assisted by the development of several rodent models that have shown that there are unique tumor-induced central and peripheral anatomical and pathophysiological changes, as well as physical and biochemical interactions between nerves, surrounding tissue, and tumor cells, which may be important to understand in order to develop better treatment strategies for cancer-associated pain. This review focuses on bone cancer pain models, nonbone cancer pain models, cancer invasion pain models, cancer chemotherapeutic-induced peripheral neuropathy, and spontaneous-occurring cancer pain models, all of which have contributed to a better understanding of the basis for tumor-induced pain and have allowed exploration of novel mechanistic-based therapies.

1. Introduction

Pain is one of many symptoms that are experienced by patients with cancer. However, if uncontrolled, it can profoundly compromise quality of life and may interfere with antineoplastic treatment (1, 2). Pain in cancer can be either constant or variable in character, owing to the inevitability that cancer involves growth and progression, changes in the tissue surrounding the tumor, and therapeutic interventions. Although no overall "cure" exists for most cancers, advances in antineoplastic therapies have allowed patients to live longer with their disease. This has raised important issues as unrelieved pain can disrupt and interfere with many routine daily activities, quality of life, and mood, thus making long-term cancer pain therapy a consideration of increasing

Chao Ma and Jun-Ming Zhang (eds.), *Animal Models of Pain*, Neuromethods, vol. 49,
DOI 10.1007/978-1-60761-880-5_8, © Springer Science+Business Media, LLC 2011

importance. Further, the incidence of cancer pain is high in patients with advanced disease (70–90%), as well as in patients undergoing active treatment for solid tumors (30–50%) (1, 3). Unfortunately, the intensity of this pain is often overwhelming. In a multisite cancer pain study, two-thirds of patients rated their worst pain during a given day a 7 out of a maximum 10, with an average pain level of 4.7 throughout the day, even though 91% of these patients were on opioid analgesics (4). Clearly, this demonstrates that most patients, even with analgesic therapy, live with moderate to severe daily pain. Cancer patients who are in advanced stages of the disease, particularly those with bone metastasis, report that they experience significant pain, and it appears that pain intensity is related to the degree of bone destruction. Similarly, pain secondary to cancer in domestic animals is a major concern in veterinary practice and should be promptly addressed to alleviate suffering, stress, and anxiety and to improve quality of life.

One complication in understanding and treating cancer pain is its variable and complex nature. In this regard, Grond et al. showed that 36% of cancer patients have neuropathic cancer pain (31% have mixed nociceptive and 5% pure neuropathic) (5). The reader is referred to Urch and Dickenson, 2008, for a mechanistic review of neuropathic pain in cancer. Collectively, these studies underscore the complex alterations that occur in the pain-sensing systems in cancer (6). Although pain due to the neoplasm is the focus of much investigation and treatment, cancer pain related to diagnostic interventions, therapeutic interventions, lumbar puncture, analgesic techniques, chemotherapy toxicity, hormonal therapy, and radiotherapy, as well as postoperative pain must also be considered and treated appropriately (7). In this regard, almost 30% of adult cancer patients and 60% of pediatric cancer patients who have undergone treatments that include radiation, chemotherapy, and/or surgery have reported experiencing pain resulting from these therapeutic procedures (8, 9).

Not only do cancer patients have to deal with persistent pain, but they also often experience "breakthrough" pain. Breakthrough or episodic pain occurs in the cancer population as a transitory pain. It occurs above tumor-induced background pain in patients, even when background pain is successfully managed by analgesics (10). Breakthrough pain is further characterized etiologically as either spontaneous or incidental. Spontaneous breakthrough pain can result from end-of-dose failure or idiopathic causes. Incidental breakthrough pain is associated with movement, whether that be of a whole limb, or something as slight as coughing. Incidental or movement-related breakthrough pain is seen in 24–93% of all patients with breakthrough pain, depending on the study (11).

Bone Metastases: Bone metastasis is the most common cause of metastatic cancer pain and the first symptom of the disease in 15–20% of patients (12). However, bone metastases are only painful in a certain number of cases, and the factors that convert a painless lesion into a painful one are not well understood (1). Pressure from the expanding tumor, release of cytokines and other chemical mediators, and ultimate fracture and tumor infiltration of surrounding nerves are generally considered to be causes of bone pain (13, 14). The pain may initially be poorly localized with a deep boring quality and unrelated to activity. In addition, it may be characterized by stabbing episodes and worsened by movement and weight bearing as the tumor progresses (15, 16).

Problems with cancer pain management: The diagnosis of the pain is not always straightforward, especially in cases of metastasis. From a mechanistic standpoint, cancer pain is often a blend of somatic, visceral, and neuropathic pain. Although this blend inevitably delays and confuses diagnosis (15), such categorization by mechanism has ultimately proven a successful approach to cancer pain management (17). With respect to treatment, concerns about the development of addiction may still lead physicians to prescribe subeffective doses of opiates (18). Conversely, subsets of patients, particularly terminal cancer patients, are often unresponsive to opioids, even at high doses. Such unresponsiveness can have several origins, including opioid tolerance, changes in opioid receptor density, higher than normal sensitivity to opioid side effects, the development of cancer-related neuropathic pain, and actual progression of the disease, that require an elevation in dose beyond the therapeutic window (19).

Analgesics in cancer pain: Pain makes a large impact on the quality of a cancer patient's life, affecting issues from mobility and independence to psychological state, and extending to include family members and other informal caregivers (20). Moreover, unrelieved pain is the symptom that people fear the most. Unfortunately, therapeutic interventions for cancer-related pain may have a negative impact in and of themselves. It is known that pain intensity varies among cancer patients and is dependent upon a patient's pain sensitivity, the type of cancer, and the tumor location (8, 21). Cancer treatment guidelines provided by the World Health Organization (WHO) have been used in oncology and pain treatment clinics (13, 22–27). Treatment of human cancer patients include the use of opioids, nonsteroidal anti-inflammatory drugs (NSAIDs), corticosteroids, local anesthetics, NMDA receptor antagonists, antidepressants, and anticonvulsants either alone or in combination. Opioids remain the "gold standard" for moderate to severe cancer pain (1, 20). However, several side effects limit opioid usefulness, including constipation, sedation, respiratory depression, vomiting (28), hallucinations, delirium, cognition failure, and, paradoxically, hyperalgesia (29), making alternative therapies attractive. Yet, opioids continue to be effective and safe analgesics in treating cancer pain; however, individual success is variable. NSAIDs play both peripheral and central roles in attenuating pain pathways; in addition, they have been used successfully to treat mild cancer pain alone or in combination with opioids for moderate pain (30) and are particularly well suited for treating pain from bone metastasis and tumor infiltration of soft tissue, myofascial tissue, or the periosteum (22). Overall management of bone cancer pain is reviewed in (31). While these medicinal treatments utilize the best drugs available at the current time, they often fail to effectively control pain in many terminal cancer patients or they have significant side effects: e.g., opioids may cause sedation, respiratory depression and interfere with gastrointestinal motility; NSAIDs may interfere with coagulation pathways or cause gastric ulcers and renal toxicity (32–37). It is clear that more effective treatments with greater efficacy for cancer pain are needed. In order to understand the basic mechanisms involved in cancer pain, and ultimately to provide insight into mechanism-based therapies, we and others have developed models of tumor-induced hyperalgesia.

2. New Approaches Using Animal Models

In spite of the need for new treatments, one of the great impediments for discovering novel analgesics is our inadequate understanding of the basic neurobiology of cancer pain generation and maintenance. Over the past two decades, a number of new animal models have been developed and used to further investigate cancer pain, neuropathic pain, and inflammatory pain (38). Study of experimental animal models has provided insight into the mechanisms that drive bone cancer pain and provides an opportunity for developing targeted therapies. In the present review, we first summarize recent findings related to the mechanisms that drive cancer pain, then briefly discuss the methods used to quantify pain in animals, and finally, describe the different types of animal models of cancer pain that have been developed. We have divided our discussion of these models into the following categories: bone cancer pain models, nonbone cancer pain models (visceral and cutaneous inoculation), cancer invasion pain models, cancer chemotherapy-related peripheral neuropathy pain models, and spontaneously occurring cancer pain models (39–41). The most commonly used animal models of cancer pain have been developed in rodents, and thus, much of this review focuses on rodent models. However, more recently described naturally occurring tumor models in dogs and cats have been developed, and these are also reviewed.

3. Literature Identification

The databases were searched using the words "cancer pain" or "tumor nociception" as the main search terms in the title, abstract, or key words of an article, and where possible, searches were performed from 1980 to 2009. Searches were carried out using the terms cancer pain or tumor nociception alone or in combination with one or more of the following words: animal models, veterinary medicine, domestic animals, rodent models, mice, rats, dogs, cats, survey, cross sectional, follow-up, prospective, longitudinal, case control, and control group. The following journals were searched by hand: Pain, Journal of Pain, European Journal of Pain, and the Journal of Cancer Pain and Symptom Palliation. Cancer sites on the Internet were also investigated for more up-to-date information on the most recent cancer pain publications. When papers were found, they were hand searched for cross-references. To avoid problems concerning the meaning and categorization of animal models of cancer pain, we included only papers in the English language.

Papers were excluded in this review if they (1) did not describe original studies, (2) did not have a well-defined control group, (3) did not include a statement indicating that the study was approved by an IACUC committee (42), or (4) were not focused on mechanisms of cancer pain. Inclusion in the review was based on the following criteria: (1) the study was well-controlled and included a well-defined control group; (2) the study involved a distinct animal model of cancer pain as opposed to models of tumor growth or metastasis; (3) the study provided mechanistic data relevant to our understanding of the causes of cancer pain.

4. Mechanisms that Drive Cancer Pain

During the past 25 years, a large number of references in the clinical literature indicated that cancer pain is generated and maintained by one of the following anatomical mechanisms: compression of bone, soft tissue, and/or peripheral nerve, vascular occlusion, and/or tumor infiltration. In addition, cancer pain can also arise as a result of diagnostic or therapeutic surgical procedures (such as biopsies and resection), or particularly in people as a side effect of toxicity relating to therapies used to treat cancer (for example, chemotherapy and radiation therapy). While the above-mentioned anatomical explanations including compression, vascular occlusion, and tumor infiltration provide a mechanistic rationale that explains the basis of tumor-induced pain in gross pathological terms, they fail to consider the *basic biochemical, molecular, and neurobiological mechanisms that underlie the production of pain*, and these factors are possibly the key elements of effective pain control in cancer patients. In this regard, tumor pain and in particular bone cancer pain represents one of the most severe types of chronic pain in both humans and animals. Unfortunately, since the mechanisms that generate cancer pain were until very recently poorly understood, the management of cancer pain has been largely empirical and based on scientific studies of noncancerous conditions such as inflammatory pain, where knowledge of the nociceptive mechanisms is quite extensive.

During the past decade, this lack of knowledge of the molecular, biochemical, and neurobiological mechanisms that generate cancer pain has begun to be addressed, with the recent development of cancer pain models (43–54), which are described in more detail below. These studies have resulted in the beginnings of a mechanism-based understanding of the factors that generate and maintain cancer-induced pain. In this regard, it is now recognized that tumor cells themselves release a number of mediators that directly affect primary afferent pain fibers. However, in addition to cancer cells, it is important to note that tumors also contain inflammatory cells and blood vessels which are often found in close proximity to primary afferent nociceptors and release mediators that affect these nociceptors. Thus, cancer cells, inflammatory cells, and vascular cells release a variety of products, including prostaglandins, ATP, bradykinin, cytokines, chemokines, nerve growth factor, and several vascular factors including endothelin 1 and vascular endothelial growth factor (VEGF), that either excite or sensitize the nociceptor. Once the nociceptor is activated, it sends an excitatory signal to the spinal cord where the nociceptive information is processed and then relayed via the spinothalamic, spinoreticulothalamic, and spinocervicothalamic tracts to higher centers of the brain. Based on the recently acquired knowledge of nociceptive mediators released at the tumor site, newer studies using animal models of cancer pain have examined blocking tumor-associated mechanisms, including endocannabinoid signaling (55), endogenous opioid systems (56, 57), TRPV1 (58–61), p38 mitogen (62), TNF-α (63, 64), endothelin (47, 65–67), IL-1 (68), CGRP (69), NGF (70), or COX-2 (71). While this information holds promise for the development of new therapeutics for the treatment of cancer pain, it is important to point out that blocking these mediators individually is not sufficient to block cancer pain completely, indicating that tumor-induced pain is produced by multifaceted mechanisms. There are several recent reviews that focus on the mechanisms that drive cancer pain and on the mediators involved, and the reader is referred to these excellent sources for more complete summaries of this information (48, 72–86).

5. Methods Used to Quantify Pain in Animals

There are a number of behavioral tests that are used to quantify pain in rats and mice, and some of these tests have recently been adopted to test pain in dogs and cats. The two most common tests are the: radiant heat paw withdrawal test developed by Hargreaves and colleagues (87) and the von Frey test (88). The radiant heat paw withdrawal test is used to assess thermal sensitivity. In this case, a noxious stimulus, a high-intensity beam from a projector lamp bulb located below an unheated glass floor, is aimed at the plantar surface of the mid hind paw. The latency in seconds to withdrawal or pain behavior (vigorous shake) is measured and recorded as a measure of tumor-induced thermal hyperalgesia or allodynia. In the von Frey test, filaments of different thicknesses are applied against the central edge of the animal hind paw. Paw withdrawal caused by the stimulation is registered as a response. There are a number of other behavioral tests that have been used to measure pain in animals, so the reader is referred to several papers and reviews that cover methods used to quantify pain in animals (40, 89–92).

It is important to point out that when measuring chronic pain conditions like tumor pain, a number of other measures have been proposed: analgesic self-administration, conditioned place aversion, gait/weight-bearing disturbance, grip/bite force, grooming (scratching/licking/biting) behavior, guarding (abnormal positioning), hind paw lifting/flinching/shaking, hypolocomotion, hypophagia, and weight loss, inattention to novel stimuli, and ultrasonic vocalization. Unfortunately, such spontaneous measures of chronic pain have only been utilized in a handful of studies employing animal models of cancer pain. For example, Betourne et al. found no place preference associated with morphine injection in mice with chronic inflammatory pain or tumor-induce hyperalgesia (57). As Mogil and Crager (2004) point out, "The greater practical demands associated with measuring spontaneous nociception in animals, combined with the lack of consensus over exactly which behavior(s) to measure, have conspired to favor the continuing and virtually exclusive measurement of hypersensitivity states" (92).

6. Models of Cancer Pain

Bone cancer produces one of the most painful conditions that exist in humans and animals. It also represents the most common pain in human patients with advanced cancer since most common tumors including breast (93), prostate, and lung have a remarkable affinity to metastasize to bone (13). Thus, it is perhaps not surprising that the first animal models developed to study cancer pain were models of primary and metastatic bone tumors. This was followed by the development of nonbone cancer pain models that mimic other types of cancers including pancreatic cancer (94, 95), squamous cell carcinoma (96), cutaneous or melanoma cancer pain (54, 57, 66), and benign, but painful, neuromas (97, 98). In addition, animal cancer pain models have been developed that replicate the pain caused by tumor invasion of peripheral nerves and the pain produced by cancer chemotherapy-related peripheral neuropathy. Finally, naturally occurring tumors that arise spontaneously in animals are being used as more natural models of cancer pain. Each of these different models will be discussed in more detail below.

7. Bone Cancer Pain Models

The most common presenting symptom of bone cancer is bone pain, and as the tumor grows, the pain becomes more severe (99). As bone pain becomes more severe, bone remodeling occurs, and as this remodeling progresses, breakthrough pain can occur during weight-bearing or movement of the affected bone, and this pain is difficult to treat with standard therapies (80). Thus, animal models that mimic both bone tumor pain and bone remodeling hold promise for understanding the mechanisms contributing to the development of tumor-induced bone pain. In this regard, one of the major advantages of recently developed animal models of bone cancer pain is that they share many characteristics that occur in the human bone cancer condition, including the pain and skeletal remodeling that accompanies metastatic bone cancer (100, 101). These bone cancer pain models are based on an intramedullary injection of cancer cells directly into bone, and thus, the location of the resulting bone tumor can be carefully controlled compared to systemic or intracardiac administration of tumor cells. In addition, these models allow easier assessment of tumor growth over time, as well as radiographic imaging, bone destruction observation, histopathological changes, accurate site-specific behavioral analyses, and appraisal of both neurochemical and neuroanatomical changes that occur at the tumor site, in the dorsal root ganglion and within the spinal cord and other levels of the central nervous system (CNS). Animal models of bone cancer pain have been developed in both mouse and rat, and these models are discussed separately. Because of their small bone size, mouse models are usually generated by surgically implanting tumor cells directly into the bone, i.e., femur or humerus. On the other hand, rat models are generally produced by percutaneous injection of cancer cells into the bone, i.e., tibia (84).

7.1. Mouse Bone Cancer Pain Models

The first animal model of cancer pain was developed in 1998 by Drs. Paul Wacnik, Al Beitz, Pat Mantyh, George Wilcox, and colleagues at the University of Minnesota (53). In this mouse model, 10^5 (20 μl) fibrosarcoma cells (NCTC 2472) were implanted directly into the femur. A crucial component of this model is that the tumor cells are confined to the marrow space of the injected femur and do not invade adjacent soft tissues (51). After injection, both ongoing and movement-evoked pain-related behaviors increase as the cancer cells proliferate and the tumor develops. These behaviors are correlated with the progressive tumor-induced bone destruction that ensues and seem to mimic those of patients with primary or metastatic bone cancer (73), this model was subsequently used to obtain new information regarding the mechanisms that generate bone cancer pain (48, 51, 53, 62, 70, 102, 103).

Thus, this femur bone cancer model was used to examine tumor-induced bone destruction, pain behaviors, and spinal cord neurochemical changes in the mouse (B6C3-Fe-*a/a* wild-type and C3H/HeJ mouse strains) (43, 51). The animals with intramedullary femur tumors showed nocifensive behaviors (vocalization and guarding of the affected limb) and mechanical allodynia (a response to nonnoxious mechanical stimuli such as light touch or palpation). This model also revealed important neurochemical changes in the spinal cord including: (1) an increase in dynorphin (a prohyperalgesic neuropeptide) expression in deep laminae of the spinal cord dorsal horn, (2) an increase in *c-fos* expression (a marker of neuronal activation) in spinal cord lamina I, and (3) an internalization of substance P (an important neurotransmitter in nociception, SP) receptors in the ipsilateral tumor-injected side of the spinal cord (43, 51). These spinal cord changes are normally found after application of a noxious stimulus; however, they are also present in cancer animals after a nonnoxious stimulus, in this case palpation.

Another finding in tumor animals was a massive astrocyte hypertrophy in the spinal cord dorsal horn but without indications of increased microglia activity. This was in contrast to what was seen in a neuropathic pain model where astrogliosis was restricted to areas of microglial activity (Hald et al. 2009). Although this massive astrocyte hypertrophy was evident in this bone cancer model, it remains unclear as to whether this increase is related to the generation and/or maintenance of bone cancer pain. These profound neurochemical changes and reorganization of the spinal cord may be involved in central sensitization (43, 51). In human cancer patients, pain seems to closely relate to the degree of bone destruction (78, 79). Since it is well accepted that osteoclast activity is crucial for bone resorption (75, 78, 79) if osteoclast activity causing bone destruction could be inhibited, cancer-induced pain should be relieved (76, 84, 104–106). Based on this concept, Honore et al. (2000) demonstrated that they could administer the novel analgesic, osteoprotegerin ligand (OPG), and successfully treat cancer pain in male C3H/HeJ mice. OPG is a member of the tumor necrosis factor (TNF) family, and when administered, it blocks osteoclast activity resulting in cancer-induced bone destruction. Their study found that OPG inhibited both the pain-related behaviors and the neurochemical changes (increased dynorphin, *c-fos* expression, internalization of SP receptors, and astrocyte hypertrophy) observed in cancer-inoculated mice (105).

Several studies have now examined the efficacy of morphine in bone cancer pain compared to inflammatory pain. In bone tumor mice, the effective morphine dose required to relieve tumor-induced pain was ten times higher than that required in mice with inflammatory pain induced by injection of complete Freund's adjuvant (CFA, an algesic agent) (107). It has subsequently been shown that this is due to a downregulation of mu opioid receptors in the spinal (dorsal root) ganglia of tumor animals compared to animals with inflammatory pain where mu opioid receptor expression is actually increased (56). In addition, King et al. reported that morphine treatment accelerates sarcoma-induced bone pain, bone loss, and spontaneous fracture by accelerating osteoclast activity and by increasing IL-1β within the femurs of implanted mice (108). Interestingly, Niiyama et al. showed that a TRPV1 antagonist potentiate the analgesic effects of systemic morphine on bone cancer pain behaviors (102).

Similarly, fibrosarcoma cells implanted into the humerus which may be more relevant to canine osteosarcoma patients has been shown to produce forelimb hyperalgesia and required a threefold increase in the morphine dose to effectively treat hyperalgesia as compared to the morphine dose needed to reduce hyperalgesia induced by the injection of carrageenan, an algesic agent, into the triceps (46, 49). Furthermore, pain-related behaviors can be effectively and dose-dependently treated with fentanyl (35), sufentanyl (35), morphine (35, 49), and anti-nerve growth factor (109). Cancer-induced hyperalgesia can also be attenuated by peripheral opioids (110) or cannabinoid receptor agonists (111). Hald and colleagues showed that repeated low dose WIN 55,212 reduced pain-related behavior and spinal changes in tumor-bearing mice without a reduction in bone destruction. WIN efficacy was not seen in neuropathic pain animals (61).

Simone and colleagues performed electrophysiological recordings from primary afferent fibers innervating a bone tumor produced by implanting fibrosarcoma (10^5 cells in 10 μl) cells into the calcaneous bone and surrounding tissue in male mice (112, 113). Tumor-implanted mice showed pain-related behaviors and mechanical hyperalgesia (an increased pain response to a mechanical stimulus that is normally painful) (112). They found that thirty-four percent of pain fibers (C fibers) in these tumor-implanted mice electrophysiologically showed a tumor-induced development of spontaneous activity and a decrease in thermal thresholds for activation, suggesting the development of activation and sensitization of C fibers (112). Khasabov et. al. (113) showed that wide dynamic range (WDR) neurons that exhibited ongoing activity and their evoked discharge rates were greater in tumor-bearing than in control mice. In addition, WDR neurons exhibited lower response thresholds for mechanical and heat stimuli, and increased responses to suprathreshold mechanical, heat, and cold stimuli (113). When fibrosarcoma or melanoma cells were implanted into the calcaneous bone of male B6C3fe/1 mice, only fibrosarcoma-injected mice developed cancer-related bone destruction and thermal hyperalgesia (an increased pain response to a normally painful thermal stimulus) (47). The peptide endothelin-1 (ET-1), which is expressed in numerous tumor types, was also found to be involved in bone cancer pain (114–117). High levels of ET-1 and activation of primary afferent fibers were observed in the fibrosarcoma tumor site, but not in a control melanoma tumor site (47, 65–67). In addition, hyperalgesia was only found in fibrosarcoma-implanted mice, suggesting that ET-1 contributes to cancer-related pain associated with fibrosarcoma tumors (47, 118, 119). This is in contrast to Fujita et al. 2008 where elevated levels of ET-1 were found, and tumor-induced hyperalgesia was observed in C57BL6 mice with subcutaneous melanoma cell inoculation. Fibrosarcoma-implanted mice also exhibited a significantly elevated level of tumor necrosis factor-α (TNF-α, a proinflammatory cytokine released by various cell types including mast cells, macrophages, endothelial cells, and fibroblasts) (64). In this study, intraplantar injection of TNF-α caused mechanical hyperalgesia in naive mice and increased hyperalgesia in fibrosarcoma-implanted mice. Similarly, Constantin et al 2008 showed TNF-α as a key player in cancer-related heat hyperalgesia and nociceptor sensitization, which generates TRPV1 upregulation and sensitization via TNFR2 in lung carcinoma implants (63).

Jasmin and colleagues from UCSF and the University of Minnesota also studied the effects of radiotherapy on tumor-induced pain (120). Sarcoma cells (2×10^5 in 5 μl) were implanted into humeri of female C3H/HeJ mice. Seven days after 6-Gy radiation, tumor-implanted mouse performance on both the rotarod and grip force tests showed significant improvement. A similar result was obtained when using the COX (cyclooxygenase) inhibitor, ketorolac. In addition, following radiation, the increased dynorphin levels and astrocyte hypertrophy seen with other mouse cancer pain models were both significantly reduced (120). When sarcoma cells (10^5 in 20 μl) were implanted into a mouse femur to examine the effects of 20-Gy and 30-Gy radiation as a means to control cancer pain, the radiation therapies with 20-Gy and 30-Gy effectively decreased cancer-induced osteolysis, reduced tumor size by 75%, and decreased bone cancer-related pain (121–123). (122). Likewise, enzastaurin, a protein kinase C beta (PKCbeta) inhibitor in combination with localized radiation treatment suppresses tumor growth and alleviated pain as compared to radiation-only treatment (124). 4T1 murine breast carcinoma cells used in a tumor xenograft bone metastases model in C3H/Scid was used to explore radiation therapy alone and in combination with antiangiogenic therapy. Results revealed that local irradiation in combination with bevacizumab enhanced radiation control of bone destruction, reduced pain-like behaviors, decreased levels of Substance P expression, and reduced tumor microvessel density when compared to single modality therapy (125). These studies demonstrated that radiation therapy effectively decreased cancer-induced pain by direct effects on tumor cells (121, 123). In addition, in mice (C3H/HeJ) with hepatocellular carcinoma (HCa-1), radiotherapy dramatically decreased CGRP expression in the spinal cord, without a change in SP level or *c-fos* expression, and this was associated with reduced responses to mechanical stimuli (126, 127).

7.2. Rat Bone Cancer Pain Models

A rat model of cancer pain using mammary gland carcinoma cells was developed in 2002 (44). In this model, either 3×10^3 or 3×10^4 MRMT-1 mammary gland carcinoma cells were implanted into female Sprague–Dawley rat tibias. Animals inoculated with MRMT-1 cells gradually showed signs of mechanical hyperalgesia in weight-bearing tests and developed mechanical allodynia using von Frey monofilments on days 10–12 and 12–14 postimplantation, respectively. In MRMT-1 rats, bone destruction was evident by day 15. The numbers of tartarate-resistant acid phosphatase positive polykaryocytes, activated by prostaglandins, cytokines, and growth factors from tumor cells, were also increased. Similar to the mouse cancer models, spinal astrocyte hypertrophy was evident by day 17 using a marker of astrocyte activity, glial fibrillary acidic protein (GFAP), and this increased astrocyte activation was also specific to bone cancer. In this study, rats implanted with 3×10^4 MRMT-1 cells were euthanized on day 16 due to bone deterioration. No significant changes were found in both heat-killed MRMT-1 or vehicle groups, and weight loss and body temperature remained unchanged in all groups. Furthermore, morphine dose-dependently attenuated mechanical allodynia and hyperalgesia, while the COX-2 inhibitor, celebrex, was ineffective, suggesting that prostaglandins may not contribute to cancer pain in the rat MRMT-1 carcinoma model (44, 52). This ineffective COX-2 treatment (52) differs from the results by Gonzalez and colleagues showing that administration of cyclooxygenase-2 (COX-2) inhibitor, lumiracoxib, administered twice daily for

10 days, attenuated mechanical hyperalgesia and bone destruction (128). In addition, COX-2 inhibitors were able to attenuate pain in a mouse model (71). Anticonvulsant drug lacosamide, was also effective in attenuating pain (129). Compared with previous bone cancer mouse models, this rat model preparation does not affect the joints, muscles, or ligaments. The use of this rat model in which 3×10^3 MRMT-1 cells are implanted into the tibia suggests that it may be a more suitable cancer pain model since both mechanical allodynia and hyperalgesia developed without significant undesirable side effects during the study's 20-day time course (44). Both of these rat and mouse models suggest that different bone cancer models may have different underlying mechanisms depending on species, time, tumor types, or tumor location.

In addition, electrophysiological recordings of MRMT-1 animals showed: (1) the receptive field size for superficial spinal cord neurons was enlarged; (2) nociceptive specific neurons in the spinal cord that normally respond to only noxious stimuli were excited by nonnoxious stimuli (45). The responses of superficial WDR neurons, generally excited by both nonnoxious and noxious stimuli, were dramatically enhanced, while deeper WDR neurons showed minimal changes, suggesting involvement of both ascending and descending facilitation pathways (37, 45). To further study descending pathway modulation, a serotonergic (5-HT_3) receptor antagonist, ondansetron, was spinally administered in MRMT-1 cancer rats (130). Ondansetron significantly decreased responses to mechanical and thermal stimuli, but not to electrical stimuli, in both tumor and naïve animals (130). In MRMT-1 cancer rats, gabapentin effectively attenuated pain-related behaviors, and electrophysiological recordings in the spinal cord indicated reduced responses to electrical and mechanical, but not thermal stimuli (131). Glial cells in the spinal cord reportedly play a role in pain perception (132–134). Therefore, Lao et al. (2005) implanted AT-3.1 prostate cancer cells (3×10^5 in 10 μl) into the tibia of male Copenhagen rats to study spinal glial activation under conditions of bone cancer pain (135). The prostate cancer animals demonstrated several characteristics: pain-related behaviors including thermal hyperalgesia, mechanical hyperlagesia, and flinches; bone destruction 1 week following tumor implantation; massive astrocyte hypertrophy in the ipsilateral side of the spinal cord; upregulation of spinal cord interleukin-1β (IL-1β, a proinflammatory cytokine) (135). Further, upregulation of IL-1β (136) and preprodynorphin mRNA and dynorphin was inhibited by electroacupuncture compared to sham control. Intrathecal injection of antiserum against dynorphin A (1–17) also significantly inhibited the cancer-induced hyperalgesia (137).

Skeletal metastasis is a serious complication of certain neoplastic diseases such as breast, prostate, and lung cancer (138, 139). To investigate this, MDA-MB-231 human breast cancer cell line

(5×10^5 in 1 ml) were injected into the femoral artery in nude rats (140). It was found that osteolytic lesions occurred exclusively in the femur, tibia, and fibula of these animals, and if the tumor cells were preincubated with an antibody against bone sialoprotein, there was a significant reduction in osteolytic lesion size. Conversely, Liepe et al. injected R-3327 prostate cancer cells directly into the left cardiac ventricle, intravenously or intraosseously into male Copenhagen rats to observe metastatic lesions and their relationship to pain (141, 142). Bone lesions were observed in bone scans following intraosseous injection, but not following intraventricular or intravenous injections of prostate cancer cells. These investigators concluded that the intraosseous administration of R-3327 prostate cancer cells represents a useful and effective osteoblastic bone lesion model.

8. Nonbone Cancer Pain Models

8.1. Visceral Pain

Visceral cancer pain is difficult to detect, and its clinical symptoms are usually unnoticed until cancer has progressed to AN advance stage. Models of pancreatic cancer pain are one of the few models used in visceral cancer pain studies (94), even though pancreatic cancer accounts for only 2% of new cancers diagnosed in the US (143). In a transgenic mouse model with spontaneous pancreatic cancer development, observations were made between wild-type (WT) and experimental (ET) groups. Pain-related behaviors, hunching and vocalization, were studied at early, mid, and late stages of cancer and the involvement of the opioid endogenous system was also examined (94, 95). The results showed: palpable cancerous nodules between weeks 9–12; 12% metastasized cancer in ET mice; increased vascular density, macrophages, CGRP (Calcitonin gene-related peptide) levels, nerve growth factor (NGF) levels, density of sensory and sympathetic fibers; precancerous cellular changes were seen as early as 6 weeks (94, 95). Pain-related behaviors, vocalization and hunching, were noticeable at the late stage, and this could be attenuated by morphine administration. However, after administration of the CNS-penetrant opioid antagonist, naloxone, but not after administration of a CNS-nonpenetrant opioid antagonist, these early stage pancreatic cancer-ET mice, demonstrated pain-related behaviors, which they normally would not exhibit at this early stage. This study revealed the following: the relationship of disease progression and pancreatic cancer pain-related behaviors is dependent on endogenous opioids. In order to reveal the presence of pancreatic pain at the early stages of tumor development, blocking the central opioid system is required; conversely administration of a central opioid agonist is required to relieve pain during the late stage of pancreatic cancer (94, 95, 144). In addition, it seems that pancreatic cancer cells that infiltrate the perineurium of local intrapancreatic nerves might cause pancreatic neuropathy (145), and therefore visceral pain. To date, we do not completely understand what causes pain in pancreatic cancer, but the generation and maintenance of pancreatic cancer-related pain may involve neurogenic inflammation (146), and there is evidence that administration of a vanilloid antagonist might be one of the effective treatments of choice (147).

8.2. Cutaneous Inoculation

In a recent study by Consantin, Lung carcinoma cells (ETCC clone1642; European Collection of Cell Cultures, Wiltshire, UK) were injected subcutaneously in the plantar and dorsal side of C57BL/6J wild-type and transgenic mice, resulting in the development of heat and mechanical sensitivity. In addition, significant elevated levels of both TNF-α and IL1β and reduced levels of IL-6 were observed in tumor homogenates. Utilizing the skin-nerve preparation and single fiber recordings, it was found that 31% (17 of 54) of the C-fibers investigated in the tumor-treated mice exhibited irregular resting activity compared with only 7% in healthy skin. In transgenic mice with a deleted (TNFR2) gene, heat hyperalgesia was attenuated and TRPV1 upregulation was prevented. In tumor bearing-mice, TNFR1 gene deletion did not result in significant differences from wild-type effects. Thus, endogenous TNF-α serves as a key player in cancer-related heat hyperalgesia and nociceptor sensitization which generates TRPV1 upregulation and sensitization via TNFR2 (63).

In a rat hind paw model, the implantation of SCC-158 cells (squamous cell carcinoma) resulted in mechanical allodynia, thermal hyperalgesia, as well as signs of spontaneous nocifensive behavior as early as 3 days postinoculation. Intraplantar administration of the TRPV1 antagonist capsazepine or TRP channels antagonist ruthenium red did not inhibit spontaneous nocifensive behavior;however, it completely inhibited mechanical allodynia and thermal hyperalgesia in tumor-bearing animals. Furthermore, an upregulation of TRPV1 receptors were found in large DRG neurons 24 days postinoculation (59).

In a model of melanoma-induced pain where B16-BL6 melanoma cells are inoculated into the hind paw of C57BL/6 mice, Sasamura et al. (2002) showed that morphine could suppress tumor growth and metastasis in mice (54). This model has been used to evaluate nonopioid analgesics. In this regard, gapapentin provided consistent inhibition of tumor-induced mechanical allodynia and hyperalgesia without analgesic tolerance (148). Intraplantar injections of the ETA receptor antagonist, BQ-123, but not the ETB-receptor antagonist, BQ-788, inhibited mechanical allodynia in tumor-bearing mice. Further, ET-1 levels in the tumor masses increased with tumor size, and mRNA of ETA, but not ETB, receptor, was significantly increased in the dorsal root ganglia on the inoculated side (66). Tsugunobu used this model to examine the analgesic efficacy and tolerance to morphine, mexiletine hydrochloride, ketamine hydrochloride (which was without effect). Morphine, mexiletine, and baclofen were all efficacious with no tolerance developing after the sixth administration of morphine and the seventh administration of mexiletine and baclofen (149). Further, this model has been used to study the reinforcing activity of morphine. It was found that neither chronic inflammatory pain nor chronic cutaneous cancer pain showed a place preference for morphine. Mice with chronic inflammation and tumor-bearing mice showed an upregulation of the NPFF system in the amygdaloid area suggesting that this area may contribute to the suppression of morphine's rewarding effect (57).

9. Cancer Invasion Pain Models

Cancer invasion of peripheral nerves often occurs in patients with vertebrae metastasis, patients with malignant lymphomas, or during tumor progression as the tumor invades surrounding nerve bundles. Each of these conditions can lead to tumor-induced neuropathic pain syndromes (150). Thus, animal models that mimic cancer-induced neuropathic pain have been developed, which can be broadly classified as cancer invasion pain models (151). In an initial study from our group, we showed that implantation of fibrosarcoma cells near the sciatic nerve produced significant mechanical allodynia 11–23 days postimplantation, which correlated with the tumor cell perineural invasion of the nerve (48). In a similar experimental design, Meth-A sarcoma cells were used to induce cancer-related nerve injury or neuropathy by implanting these cells in close proximity to the sciatic nerve in male BALB/c mice. This model benefited from the slow progression of this rather aggressive tumor (151), but more importantly, it illustrated that cancer-related neuropathy causes spontaneous pain (paw lifting and guarding) and both thermal hyperalgesia and allodynia. This is consistent with the increased pain that is associated with tumor invasion of nerves in human patients (152). Mechanical allodynia was also visible on day 10 but changed into mechanical hyposensitivity on day 14. Damage to both myelinated and nonmyelinated fibers were found to be more extensive in this cancer-induced neuropathy model than in the sciatic nerve ligation model (chronic constriction injury (CCI)), suggesting cancer-associated nerve compression differs from nerve ligation. Similar to other previous mouse cancer models, Meth-A sarcoma animals showed an upregulation of dynorphin, c-*fos* expression, and SP in the spinal cord (50).

10. Cancer Chemotherapy-Related Peripheral Neuropathy Pain Models

The advance technologies of treating cancer continue to help cancer patients live longer. However, cancer-induced pain is still a major problem, and major sources of cancer-induced pain are from cancer itself and therapeutic treatments such as radiation therapy, surgery, and chemotherapy (153). This type of pain can impede cancer patients' quality of life during the course of cancer treatments and cause dose limiting of the treatments. This chemotherapy-related neuropathy may continue after chemotherapeutic discontinuation (coasting) (154). This chemotherapy-induced neuropathy varies depending on dose, treatment duration, and other concurrent or preexisting conditions of the patients. The underlying cause of this chemotherapy-induced neuropathy, however, remains poorly understood, and therefore, the fourth pain model, a cancer chemotherapy-related peripheral neuropathy pain model, has been developed to understand mechanistic-based pathophysiology of chemotherpeutic agent-induced neuropathy. Using these models, cancer chemotherapeutic drugs including vincristine (155, 156), paclitaxel (157–159), and cisplatin (157, 160) yield neuropathic pain after extended use (161), and the review focuses on these three chemotherapeutic agents.

10.1. Vincristine-Related Peripheral Neuropathy Pain Models

One of the most commonly used chemotherapeutic agents is vincristine, which belongs to the vinca alkaloid family (162). It binds to intracellular tubulin and alters microtubuli structures causing dose-dependent neuropathy (154). The signs of neuropathy start with paraesthesia followed by hyperesthesia (154). A dose of vincristine as low as 50 μg/kg produces mechanical hyperalgesia, allodynia, and thermal hypoalgesia in rats (163). An intravenous vincristine injection induced mechanical hyperalgesia within 2 weeks following initiation of the chemotherapy regimes in Sprague–Dawley rats (155). However, 2 weeks following vincristine discontinuation, the signs of mechanical hyperalgesia ablated (155). Interestingly, vincristine also caused greater mechanical hyperalgesia in female vs. male rats (164). Furthermore, in vincristine-treated rats, electrophysiological responses to suprathreshold stimuli in C fiber nociceptors were greater; both C and A fiber mean conduction velocities were slower; no histopathological changes were evident (156, 165). Vincristine, therefore, does not impair nociceptor function, but enhances the mechanisms associated with suprathreshold stimuli (165). A comprehensive ultrastructural study of vincristine-induce neuropathy in the rat revealed a significant increase in the cross-sectional area of myelinated axons as well as a dramatic decrease in axonal microtubules and disorganization of microtubules in myelinated axons (166).

When vincristine was administered continuously, it dose-dependently produced mechanical allodynia, but not thermal hyperalgesia (167). Mehcanical allodynia started after 1 week of vincristine infusion and returned to the baseline values 4 weeks, while cold allodynia occurred 1 week after the infusion (168). This mechanical hyperalgesia could be attenuated by morphine or lidocaine administration (167), but not by μ opioid antagonists (155). Xiao Wh and Bennett GJ, 2007, showed significant increase in spontaneously discharging A-fibers and C-fibers following paclitaxel- and vincristine-evoked pain. In addition, prophylactic treatment with acetyl-l-carnitine, which blocks the development of the paclitaxel-evoked pain, causes a significant decrease (ca. 50%) in the incidence of A-fibers and C-fibers with spontaneous discharge. Because vincristine treatment produces different results, such as hyperalgesia, hypoalgesia, and allodynia, its mechanisms remain to be identified.

10.2. Paclitaxel-Related Peripheral Neuropathy Pain Models

Another chemotherapeutic drug causing neuropathy is paclitaxel, which is used to treat a variety of cancers: breast cancers, non-small cell lung cancers, head and neck cancers, melanoma, and ovarian cancers (169–175). A significant number of cancer patients who are treated with paclitaxel complain of tingling, numbness, and burning pain (158, 169–171, 176–181). Paclitaxel, which is a vinca alkaloid, binds to tubulin blocking the polymerization of microtubules (169) and interfering with mitosis, and it has been reported to have serious side effects, including myelosuppression and sensory or sensory autonomic neuropathy (154, 169, 176, 182–184). Paclitaxel-induced neuropathy lasted for several weeks and was mostly limited to peripheral nerves with the animals showing no clinical systemic toxicity (158, 176). In addition, this agent induced mechanical hyperalgesia and thermal hyperalgesia in Sprague–Dawley rats without motor deficits (161, 176). In another study paclitaxel was shown to impair rotarod performance during both light and dark cycles which is indicative of motor neuropathy. In this study paclitaxel caused gait disturbance as early as 2 weeks after treatment began (185). Electrophysiological recording showed a decrease in H-wave amplitudes in the hindlimb and a decrease in action potentials of sensory nerves in the tail. It affected all sensory modalities, especially those associated with thick myelinated nerve fibers (154, 186) but these could be blocked by an intraperitoneal injection of gabapentin (187). Paclitaxel-treated animals also demonstrated severe axonal degeneration and hypomyelination of dorsal roots, but not ventral roots (186). In addition, in a study of 10 different mouse strains intraperitoneal paclitaxel induced mechanical allodynia in all ten strain, with especially robust changes in the DBA/2J mouse strain (188). However, both the paclitaxel-sensitive DBA/2J strain and the resistant C57BL/6J strain showed comparable cold allodynia, but neither of them showed heat hyperalgesia (188). This chemotherapy-induced neuropathic pain mouse models demonstrated a genetic component to mechanical allodynia (188). Paclitaxel in CD1 mice also caused a decrease in peptide neurotransmitters, for instance, SP in dorsal root ganglion (DRG) and reduced action potential amplitude of the caudal nerve, which was prevented by nerve growth factor (NGF) administration (189). Glutamate may also have a neuroprotective effect in preventing neuropathy induced by paclitaxel (185).

10.3. Cisplatin-Related Peripheral Neuropathy Pain Models

The last commonly used chemotherapeutic drug for treating ovarian cancers and small cell lung cancer is cisplatin. Cisplatin causes not only ototoxicity and nephrotoxcity but also neurotoxicity such as peripheral polyneuropathy, mechanical hyperalgesia, and allodynia in rats (190). These neuropathy symptoms describe as numbness and tingling, and these symptoms can be severe with increasing cumulative doses (191). Polyneuropathy caused by cisplatin can last over 10 years, and this neuropathy depends on dose and duration used (160). Rats receving three cisplatin intraperitoneal administrations at a cumulative dose of 15 mg/kg showed mechanical allodynia and hyperalgesia which lasted up to 15 days after injection (190), and it caused gait disturbance by 8 weeks (185). One advantage of this model is the conservation of a good clinical status and rapid progression of cisplatin-induced sensory peripheral neuropathy symptoms (190). Impairment of rotarod analyses was observed only in the dark cycles, suggesting some proprioceptive loss, and these side effects could be protected by glutamate (185). Electrophysiological recording revealed a significant reduction of sensory nerve conduction velocity, but motor nerve conduction velocity remained unaffected (192). In addition, Cata and colleagues showed in Cisplatin-induced chemoneuropathy that wide dynamic range neurons had significantly higher spontaneous activity and longer after discharges to noxious mechanical stimuli and longer after discharges and abnormal windup to transcutaneous electrical stimuli (193). Cisplatin histologically affects large axons with normal myelin levels but has no effects on nonmyelinated axons. In addition, DRG apoptosis (cell death) may partly contribute to cisplatin neurotoxicity, which can be blocked by a high dose of NGF, indicating that cisplatin induced apoptosis occurs through mitochondrial stress pathways (194, 195). This neuropathy could be blocked by neurotrophic factor ACTH (4–9) analog ORG 2766 (191, 192, 196) or glial growth factor rhGGF2 (196). The survival of the large fiber sensory neurons can also be induced by a neurotrophin-3 (NT-3) (197). Although cisplatin-induced neuropathy progressed for 6 weeks after discontinuation and slowly reversed over 3 months, the development of motor neuropathy symptoms could be prevented by early decompressive surgery (198). Recently, the neuroprotective effect of IL-6 was examined in three Chemotherapy-related Peripheral Neuropathy Pain Models (vinvcristine, paclitaxel, and cisplatin). Chronic IL-6 treatment prevented the behavioral and electrophysiological abnormalities produced by vincristine, paclitaxel, and cisplatin and similarly prevented the pathological changes in peripheral nerves (199).

11. Spontaneously Occurring Cancer Pain Models

A more natural model of cancer pain involves using animals that have spontaneously occurring tumors. Such models have been used recently to evaluate better therapeutic approaches to treat cancer pain. In this regard, a spontaneous occurring osteosarcoma canine model has been used to examine the effectiveness of targeting specific nociceptive neurons in DRG as a novel method to treat bone cancer pain (60, 200), as ablating the DRG neurons expressing TRPV1 receptors might be an effective treatment for controlling chronic cancer pain while leaving other sensory functions intact. The potent TRPV1 agonist, resiniferatoxin (RTX, a capsaicin analogue), was intrathecally administered to target only the TRPV1 receptor expressing axons in the spinal cord and their cell bodies in the DRGs, while leaving other DRG neurons unaffected. When intrathecally administered, RTX kills nociceptive DRG neurons containing the TRPV1 receptor and produces a prolonged analgesic effect in spontaneous bone cancer dogs that lasts for up to 14 weeks (40). The concept of selective DRG neuron ablation to treat cancer pain is intriguing. Some laboratories have also used this concept to target other neurons, for example, targeting neurons expressing SP receptors. Such approaches have shown no serious motor or other side effects in either rat or dog models an thus represent novel cancer pain treatments (201, 202).

12. Conclusions

These experimental animal models are extremely useful representations of human tumor-induced pain and thus allow the dissection of the molecular and cellular mechanisms contributing to cancer pain. Furthermore, the use of cancer pain models has provided insights into the mechanisms driving cancer pain and has offered an opportunity for developing targeted therapy. These animal models of cancer pain suggest that different bone cancer models have different underlying nociceptive mechanisms depending on species, time, tumor types, or tumor location. In addition, because the development of cancer pain is a dynamic process, alleviating cancer pain based on disease progression may be more effective than simply administering analgesic drugs at the late stages of the disease process. At the beginning of tumor growth, as tumors start to proliferate, pronociceptive factors, such as PGE_2 and ET, are released. Therefore, drugs such as COX inhibitors (203, 204) or ET antagonists (48, 65–67, 117) may be effective treatments during this early period. Depending on the tumor type, cytokines and chemokines are released either from the tumor cells themselves or from the infiltrated immune cells (205). Since many cytokines and chemokines can directly affect primary afferent pain fibers at the tumor site, knowledge of the cytokines released by various tumor types enable the development of tumor-specific cancer pain therapies. Moreover, growing tumors often compress surrounding nerve bundles, and at this point, neuropathic pain medications may provide better analgesia. At later stages of bone tumor growth, osteoclasts typically proliferate, and in this case, medications that block osteoclast activity may be appropriate. For example, OPG or bisphosphonates (206) may yield effective pain attenuation (207–209). When tumors fill the intramedullary canal and some tumor cells start to die producing an acidic environment, TRPV1 or ASIC receptor antagonists may be more advantageous in controlling cancer pain. As bone destruction becomes evident, ATP antagonists may block movement-related pain (78, 79, 210).Such information will be critical in developing novel therapeutic drugs that specifically target certain genes for specific types of cancer pain (211).

The development of these cancer pain animal models has come at a time when cancer patients are surviving longer, so cancer pain has become a significant quality of life issue. Use of these models has shown a number of unique features that are associated with pain-related behaviors and have increased our basic tumor pain knowledge in terms of anatomy, neurobiology, neurophysiology, genetics, psychology, pharmacology, and molecular mechanisms that underlie cancer pain. Many of the features observed in these animal models are shared by human cancer patients suffering from tumor pain. Some of these shared features include bone destruction, primary afferent neuron sensitization, and the reorganization and development of central sensitization in the spinal cord. Further, the newly developed animal models have begun to provide a platform for testing new drug combinations and discerning mechanisms of interaction specific to the context of tumor pain. Although these studies provide data that allow clinicians to consider novel cancer pain therapy options, the translational highway from a single experimental animal study to the clinical evidence of the efficacy and safety of different analgesics and administration schemes in treating pain caused by tumors is both long and tortuous (212). However, this increased understanding of cancer pain mechanisms will undoubtedly lead to the development of novel therapeutic approaches, such as the use of antibodies, small interfering RNA and aptamers (213), and mechanism-based pharmacotherapeutic treatments to treat tumor-induced nociception.

References

1. Portenoy RK, Lesage P. Management of cancer pain. Lancet 1999;353:1695–700.
2. Weinstein SM. New pharmacological strategies in the management of cancer pain. Cancer Invest 1998;16:94–101.
3. Foley KM. Controlling cancer pain. Hosp Pract (Minneap) 2000;35:101, 8, 111–2.
4. Caraceni A, Portenoy RK. An international survey of cancer pain characteristics and syndromes. IASP Task Force on Cancer Pain. International Association for the Study of Pain. Pain 1999;82:263–74.
5. Grond S, Radbruch L, Meuser T, Sabatowski R, Loick G, Lehmann KA. Assessment and treatment of neuropathic cancer pain following WHO guidelines. Pain 1999;79:15–20.
6. Urch CE, Dickenson AH. Neuropathic pain in cancer. Eur J Cancer 2008;44:1091–6.
7. Cherny NI, Hiller N. Diagnostic investigations in cancer pain. Cancer Invest 2000; 18:651–63.
8. Foley KM. Advances in cancer pain. Arch Neurol 1999;56:413–7.
9. Vecht CJ. Cancer pain: a neurological perspective. Curr Opin Neurol 2000;13: 649–53.
10. Portenoy RK, Hagen NA. Breakthrough pain: definition, prevalence and characteristics. Pain 1990;41:273–81.
11. Patt RB, Ellison NM. Breakthrough pain in cancer patients: characteristics, prevalence, and treatment. Oncology (Williston Park) 1998;12:1035–46; discussion 1049–52.
12. Banning A, Sjogren P, Henriksen H. Treatment outcome in a multidisciplinary cancer pain clinic. Pain 1991;47:129–34; discussion 127–8.
13. Mercadante S. Malignant bone pain: pathophysiology and treatment. Pain 1997; 69:1–18.
14. Rubens RD. Bone metastases – the clinical problem. Eur J Cancer 1998;34:210–3.
15. Twycross RG. Management of pain in skeletal metastases. Clin Orthop Relat Res 1995;312:187–96.
16. Reale C, Turkiewicz AM, Reale CA. Antalgic treatment of pain associated with bone metastases. Crit Rev Oncol Hematol 2001; 37:1–11.
17. Ashby MA, Fleming BG, Brooksbank M, et al. Description of a mechanistic approach to pain management in advanced cancer. Preliminary report. Pain 1992;51:153–61.
18. Vilensky W. Opioid "mythstakes": opioid analgesics – current clinical and regulatory perspectives. J Am Osteopath Assoc 2002; 102:S11–4.
19. Mercadante S, Portenoy RK. Opioid poorly-responsive cancer pain. Part 1: clinical considerations. J Pain Symptom Manage 2001;21:144–50.
20. Ahmedzai S. Current strategies for pain control. Ann Oncol 1997;8 Suppl 3:S21–4.
21. Foley KM. Controlling the pain of cancer. Sci Am 1996;275:164–5.
22. Levy MH. Pharmacologic treatment of cancer pain. N Engl J Med 1996;335:1124–32.
23. Cleary JF. Cancer pain management. Cancer Control 2000;7:120–31.
24. Miguel R. Interventional treatment of cancer pain: the fourth step in the World Health Organization analgesic ladder? Cancer Control 2000;7:149–56.
25. Thomas EM, Weiss SM. Nonpharmacological interventions with chronic cancer pain in adults. Cancer Control 2000;7:157–64.
26. Bruera E, Lawlor P. Cancer pain management. Acta Anaesthesiol Scand 1997;41: 146–53.
27. Mishra S, Bhatnagar S, Singhal AK. Recent trends in cancer pain management. Indian J Med Paediatr Oncol 2004;25:22–8.
28. Chang HM. Cancer pain management. Med Clin North Am 1999;83:711–36, vii.
29. Daeninck PJ, Bruera E. Opioid use in cancer pain. Is a more liberal approach enhancing toxicity? Acta Anaesthesiol Scand 1999;43:924–38.
30. Jenkins CA, Bruera E. Nonsteroidal anti-inflammatory drugs as adjuvant analgesics in cancer patients. Palliat Med 1999;13: 183–96.
31. Wilkinson AN, Viola R, Brundage MD. Managing skeletal related events resulting from bone metastases. BMJ 2008;337: a2041.
32. Saito O, Aoe T, Yamamoto T. Analgesic effects of nonsteroidal antiinflammatory drugs, acetaminophen, and morphine in a mouse model of bone cancer pain. J Anesth 2005;19:218–24.
33. Lees P, Landoni MF, Giraudel J, Toutain PL. Pharmacodynamics and pharmacokinetics of nonsteroidal anti-inflammatory drugs in species of veterinary interest. J Vet Pharmacol Ther 2004;27:479–90.

34. Giraudel JM, Diquelou A, Lees P, Toutain PL. Development and validation of a new model of inflammation in the cat and selection of surrogate endpoints for testing antiinflammatory drugs. J Vet Pharmacol Ther 2005;28:275–85.
35. El Mouedden M, Meert TF. Evaluation of pain-related behavior, bone destruction and effectiveness of fentanyl, sufentanil, and morphine in a murine model of cancer pain. Pharmacol Biochem Behav 2005;82:109–19.
36. KuKanich B, Lascelles BD, Papich MG. Pharmacokinetics of morphine and plasma concentrations of morphine-6-glucuronide following morphine administration to dogs. J Vet Pharmacol Ther 2005;28:371–6.
37. Urch CE, Donovan-Rodriguez T, Gordon-Williams R, Bee LA, Dickenson AH. Efficacy of chronic morphine in a rat model of cancer-induced bone pain: behavior and in dorsal horn pathophysiology. J Pain 2005;6:837–45.
38. Ma C. Animal models of pain. Int Anesthesiol Clin 2007;45:121–31.
39. Pacharinsak C, Beitz A. Animal models of cancer pain. Comp Med 2008;58:220–33.
40. Brown DC, Iadarola MJ, Perkowski SZ, et al. Physiologic and antinociceptive effects of intrathecal resiniferatoxin in a canine bone cancer model. Anesthesiology 2005;103:1052–9.
41. Wang LX, Wang ZJ. Animal and cellular models of chronic pain. Adv Drug Deliv Rev 2003;55:949–65.
42. Gebhart GF. Animal models of pain. ILAR 1999;40 http://dels-old.nas.edu/ilar_n/ilarjournal/40_3/index.shtml.
43. Honore P, Schwei J, Rogers SD, et al. Cellular and neurochemical remodeling of the spinal cord in bone cancer pain. Prog Brain Res 2000;129:389–97.
44. Medhurst SJ, Walker K, Bowes M, et al. A rat model of bone cancer pain. Pain 2002;96:129–40.
45. Urch CE, Donovan-Rodriguez T, Dickenson AH. Alterations in dorsal horn neurones in a rat model of cancer-induced bone pain. Pain 2003;106:347–56.
46. Wacnik PW, Kehl LJ, Trempe TM, Ramnaraine ML, Beitz AJ, Wilcox GL. Tumor implantation in mouse humerus evokes movement-related hyperalgesia exceeding that evoked by intramuscular carrageenan. Pain 2003;101:175–86.
47. Wacnik PW, Eikmeier LJ, Ruggles TR, et al. Functional interactions between tumor and peripheral nerve: morphology, algogen identification, and behavioral characterization of a new murine model of cancer pain. J Neurosci 2001;21:9355–66.
48. Wacnik PW, Wilcox GL, Clohisy DR, Ramnaraine ML, Eikmeier LJ, Beitz AJ. Cancer pain mechanisms and animal models of cancer pain. Proceeding of the 9th World Congress on Pain, Progress in Pain Research and management 2000;16:615–37.
49. Vermeirsch H, Nuydens RM, Salmon PL, Meert TF. Bone cancer pain model in mice: evaluation of pain behavior, bone destruction and morphine sensitivity. Pharmacol Biochem Behav 2004;79:243–51.
50. Shimoyama M, Tatsuoka H, Ohtori S, Tanaka K, Shimoyama N. Change of dorsal horn neurochemistry in a mouse model of neuropathic cancer pain. Pain 2005;114:221–30.
51. Schwei MJ, Honore P, Rogers SD, et al. Neurochemical and cellular reorganization of the spinal cord in a murine model of bone cancer pain. J Neurosci 1999;19:10886–97.
52. Walker K, Medhurst SJ, Kidd BL, et al. Disease modifying and anti-nociceptive effects of the bisphosphonate, zoledronic acid in a model of bone cancer pain. Pain 2002;100:219–29.
53. Wacnik PW, Ramnaraine MLR, Stone LS, Kitto KF, Laughlin TM, Mantyh PW, Beitz AJ, Wilcox GL. A practical model of cancer pain in mice comparing hind limb implantation of melanoma cells in the sciatic sheath, intramusculary, in the medullary canal or subcutaneously.. Soc Neurosci Abstr (Los Angeles, CA) 1998;24(1):628.
54. Sasamura T, Nakamura S, Iida Y, et al. Morphine analgesia suppresses tumor growth and metastasis in a mouse model of cancer pain produced by orthotopic tumor inoculation. Eur J Pharmacol 2002;441:185–91.
55. Khasabova IA, Khasabov SG, Harding-Rose C, et al. A decrease in anandamide signaling contributes to the maintenance of cutaneous mechanical hyperalgesia in a model of bone cancer pain. J Neurosci 2008;28:11141–52.
56. Yamamoto J, Kawamata T, Niiyama Y, Omote K, Namiki A. Down-regulation of mu opioid receptor expression within distinct subpopulations of dorsal root ganglion neurons in a murine model of bone cancer pain. Neuroscience 2008;151:843–53.
57. Betourne A, Familiades J, Lacassagne L, et al. Decreased motivational properties of morphine in mouse models of cancerous- or inflammatory-chronic pain: implication of supraspinal neuropeptide FF(2) receptors. Neuroscience 2008;157:12–21.

58. Niiyama Y, Kawamata T, Yamamoto J, Omote K, Namiki A. Bone cancer increases transient receptor potential vanilloid subfamily 1 expression within distinct subpopulations of dorsal root ganglion neurons. Neuroscience 2007;148:560–72.
59. Shinoda M, Ogino A, Ozaki N, et al. Involvement of TRPV1 in nociceptive behavior in a rat model of cancer pain. J Pain 2008;9:687–99.
60. Ghilardi JR, Rohrich H, Lindsay TH, et al. Selective blockade of the capsaicin receptor TRPV1 attenuates bone cancer pain. J Neurosci 2005;25:3126–31.
61. Hald A, Ding M, Egerod K, et al. Differential effects of repeated low dose treatment with the cannabinoid agonist WIN 55,212-2 in experimental models of bone cancer pain and neuropathic pain. Pharmacol Biochem Behav 2008;91:38–46.
62. Svensson CI, Medicherla S, Malkmus S, et al. Role of p38 mitogen activated protein kinase in a model of osteosarcoma-induced pain. Pharmacol Biochem Behav 2008;90: 664–75.
63. Constantin CE, Mair N, Sailer CA, et al. Endogenous tumor necrosis factor alpha (TNFalpha) requires TNF receptor type 2 to generate heat hyperalgesia in a mouse cancer model. J Neurosci 2008;28:5072–81.
64. Wacnik PW, Eikmeier LJ, Simone DA, Wilcox GL, Beitz AJ. Nociceptive characteristics of tumor necrosis factor-alpha in naive and tumor-bearing mice. Neuroscience 2005;132: 479–91.
65. Pickering V, Jay Gupta R, Quang P, Jordan RC, Schmidt BL. Effect of peripheral endothelin-1 concentration on carcinoma-induced pain in mice. Eur J Pain 2008;12:293–300.
66. Fujita M, Andoh T, Saiki I, Kuraishi Y. Involvement of endothelin and ET(A) endothelin receptor in mechanical allodynia in mice given orthotopic melanoma inoculation. J Pharmacol Sci 2008;106:257–63.
67. Hamamoto DT, Khasabov SG, Cain DM, Simone DA. Tumor-evoked sensitization of C nociceptors: a role for endothelin. J Neurophysiol 2008;100:2300–11.
68. Baamonde A, Curto-Reyes V, Juarez L, Meana A, Hidalgo A, Menendez L. Antihyperalgesic effects induced by the IL-1 receptor antagonist anakinra and increased IL-1beta levels in inflamed and osteosarcoma-bearing mice. Life Sci 2007;81:673–82.
69. Wacnik PW, Baker CM, Herron MJ, et al. Tumor-induced mechanical hyperalgesia involves CGRP receptors and altered innervation and vascularization of DsRed2 fluorescent hindpaw tumors. Pain 2005;115: 95–106.
70. Sevcik MA, Ghilardi JR, Peters CM, et al. Anti-NGF therapy profoundly reduces bone cancer pain and the accompanying increase in markers of peripheral and central sensitization. Pain 2005;115:128–41.
71. Sabino MA, Ghilardi JR, Jongen JL, et al. Simultaneous reduction in cancer pain, bone destruction, and tumor growth by selective inhibition of cyclooxygenase-2. Cancer Res 2002;62:7343–9.
72. Mantyh PW. Cancer pain and its impact on diagnosis, survival and quality of life. Nat Rev Neurosci 2006;7:797–809.
73. Mantyh PW, Clohisy DR, Koltzenburg M, Hunt SP. Molecular mechanisms of cancer pain. Nat Rev Cancer 2002;2:201–9.
74. Donovan-Rodriguez T, Dickenson AH, Urch CE. Superficial dorsal horn neuronal responses and the emergence of behavioural hyperalgesia in a rat model of cancer-induced bone pain. Neurosci Lett 2004;360:29–32.
75. Clohisy DR, Mantyh PW. Bone cancer pain. Clin Orthop Relat Res 2003;415 (Suppl):S279–88.
76. Honore P, Mantyh PW. Bone cancer pain: from mechanism to model to therapy. Pain Med 2000;1:303–9.
77. Urch CE. Current and potential treatment for cancer-induced bone pain. *AAHPM* 2004;5(4):1–3, 7.
78. Mantyh PW, Hunt SP. Mechanisms that generate and maintain bone cancer pain. Novartis Found Symp 2004;260:221–38; discussion 238–40, 277–9.
79. Mantyh PW. A mechanism-based understanding of bone cancer pain. Novartis Found Symp 2004;261:194–214; discussion 214–9, 256–61.
80. Mercadante S, Arcuri E. Breakthrough pain in cancer patients: pathophysiology and treatment. Cancer Treat Rev 1998;24:425–32.
81. Regan JM, Peng P. Neurophysiology of cancer pain. Cancer Control 2000;7:111–.
82. Sabino MA, Mantyh PW. Pathophysiology of bone cancer pain. J Support Oncol 2005; 3:15–24.
83. Urch C. The pathophysiology of cancer-induced bone pain: current understanding. Palliat Med 2004;18:267–74.
84. Goblirsch MJ, Zwolak P, Clohisy DR. Advances in understanding bone cancer pain. J Cell Biochem 2005;96:682–8.
85. Goblirsch MJ, Zwolak PP, Clohisy DR. Biology of bone cancer pain. Clin Cancer Res 2006;12:6231s–5s.

86. Gordon-Williams RM, Dickenson AH. Central neuronal mechanisms in cancer-induced bone pain. Curr Opin Support Palliat Care 2007; 1:6–10.
87. Hargreaves K, Dubner R, Brown F, Flores C, Joris J. A new and sensitive method for measuring thermal nociception in cutaneous hyperalgesia. Pain 1988;32:77–88.
88. von Frey M. Verspätete schmerzempfindungen. Z gesamte Neurol Psychiat 1922;79: 324–33.
89. Jourdan D, Ardid D, Eschalier A. Automated behavioural analysis in animal pain studies. Pharmacol Res 2001;43:103–10.
90. Lariviere WR, Wilson SG, Laughlin TM, et al. Heritability of nociception. III. Genetic relationships among commonly used assays of nociception and hypersensitivity. Pain 2002;97:75–86.
91. Mogil JS, Wilson SG, Bon K, et al. Heritability of nociception I: responses of 11 inbred mouse strains on 12 measures of nociception. Pain 1999;80:67–82.
92. Mogil JS, Crager SE. What should we be measuring in behavioral studies of chronic pain in animals? Pain 2004;112:12–5.
93. Guise TA, Kozlow WM, Heras-Herzig A, Padalecki SS, Yin JJ, Chirgwin JM. Molecular mechanisms of breast cancer metastases to bone. Clin Breast Cancer 2005;5 Suppl: S46–53.
94. Lindsay TH, Jonas BM, Sevcik MA, et al. Pancreatic cancer pain and its correlation with changes in tumor vasculature, macrophage infiltration, neuronal innervation, body weight and disease progression. Pain 2005;119:233–46.
95. Sevcik MA, Jonas BM, Lindsay TH, et al. Endogenous opioids inhibit early-stage pancreatic pain in a mouse model of pancreatic cancer. Gastroenterology 2006;131:900–10.
96. Nagamine K, Ozaki N, Shinoda M, et al. Mechanical allodynia and thermal hyperalgesia induced by experimental squamous cell carcinoma of the lower gingiva in rats. J Pain 2006;7:659–70.
97. Dorsi MJ, Chen L, Murinson BB, Pogatzki-Zahn EM, Meyer RA, Belzberg AJ. The tibial neuroma transposition (TNT) model of neuroma pain and hyperalgesia. Pain 2008; 134:320–34.
98. Tyner TR, Parks N, Faria S, et al. Effects of collagen nerve guide on neuroma formation and neuropathic pain in a rat model. Am J Surg 2007;193:e1–6.
99. Coleman RE. Metastatic bone disease: clinical features, pathophysiology and treatment strategies. Cancer Treat Rev 2001;27: 165–76.
100. Coleman RE, Rubens RD. The clinical course of bone metastases from breast cancer. Br J Cancer 1987;55:61–6.
101. Mercadante S, Fulfaro F. Management of painful bone metastases. Curr Opin Oncol 2007;19:308–14.
102. Niiyama Y, Kawamata T, Yamamoto J, Furuse S, Namiki A. SB366791, a TRPV1 antagonist, potentiates analgesic effects of systemic morphine in a murine model of bone cancer pain. Br J Anaesth 2009;102:251–8.
103. Brainin-Mattos J, Smith ND, Malkmus S, et al. Cancer-related bone pain is attenuated by a systemically available delta-opioid receptor agonist. Pain 2006;122:174–81.
104. Clohisy DR, Ramnaraine ML. Osteoclasts are required for bone tumors to grow and destroy bone. J Orthop Res 1998;16: 660–6.
105. Honore P, Luger NM, Sabino MA, et al. Osteoprotegerin blocks bone cancer-induced skeletal destruction, skeletal pain and pain-related neurochemical reorganization of the spinal cord. Nat Med 2000;6:521–8.
106. Thompson SW, Tonge D. Bone cancer gain without the pain. Nat Med 2000;6:504–5.
107. Luger NM, Sabino MA, Schwei MJ, et al. Efficacy of systemic morphine suggests a fundamental difference in the mechanisms that generate bone cancer vs inflammatory pain. Pain 2002;99:397–406.
108. King T, Vardanyan A, Majuta L, et al. Morphine treatment accelerates sarcoma-induced bone pain, bone loss, and spontaneous fracture in a murine model of bone cancer. Pain 2007;132:154–68.
109. Halvorson KG, Kubota K, Sevcik MA, et al. A blocking antibody to nerve growth factor attenuates skeletal pain induced by prostate tumor cells growing in bone. Cancer Res 2005;65:9426–35.
110. Menendez L, Lastra A, Hidalgo A, Meana A, Garcia E, Baamonde A. Peripheral opioids act as analgesics in bone cancer pain in mice. Neuroreport 2003;14:867–9.
111. Kehl LJ, Hamamoto DT, Wacnik PW, et al. A cannabinoid agonist differentially attenuates deep tissue hyperalgesia in animal models of cancer and inflammatory muscle pain. Pain 2003;103:175–86.
112. Cain DM, Wacnik PW, Eikmeier L, Beitz A, Wilcox GL, Simone DA. Functional interactions between tumor and peripheral nerve in a model of cancer pain in the mouse. Pain Med 2001;2:15–23.

113. Khasabov SG, Hamamoto DT, Harding-Rose C, Simone DA. Tumor-evoked hyperalgesia and sensitization of nociceptive dorsal horn neurons in a murine model of cancer pain. Brain Res 2007;1180:7–19.
114. Kurbel S, Kurbel B, Kovacic D, et al. Endothelin-secreting tumors and the idea of the pseudoectopic hormone secretion in tumors. Med Hypotheses 1999;52:329–33.
115. Asham E, Shankar A, Loizidou M, et al. Increased endothelin-1 in colorectal cancer and reduction of tumour growth by ET(A) receptor antagonism. Br J Cancer 2001;85:1759–63.
116. Zhou QL, Strichartz G, Davar G. Endothelin-1 activates ET(A) receptors to increase intracellular calcium in model sensory neurons. Neuroreport 2001;12:3853–7.
117. Davar G. Endothelin-1 and metastatic cancer pain. Pain Med 2001;2:24–7.
118. Peters CM, Lindsay TH, Pomonis JD, et al. Endothelin and the tumorigenic component of bone cancer pain. Neuroscience 2004; 126:1043–52.
119. Hasue F, Kuwaki T, Yamada H, Fukuda Y, Shimoyama M. Inhibitory actions of endothelin-1 on pain processing. J Cardiovasc Pharmacol 2004;44:S318–20.
120. Vit JP, Ohara PT, Tien DA, et al. The analgesic effect of low dose focal irradiation in a mouse model of bone cancer is associated with spinal changes in neuro-mediators of nociception. Pain 2006;120:188–201.
121. Goblirsch M, Mathews W, Lynch C, et al. Radiation treatment decreases bone cancer pain, osteolysis and tumor size. Radiat Res 2004;161:228–34.
122. Goblirsch M, Zwolak P, Ramnaraine ML, et al. Novel cytosine deaminase fusion gene enhances the effect of radiation on breast cancer in bone by reducing tumor burden, osteolysis, and skeletal fracture. Clin Cancer Res 2006;12:3168–76.
123. Goblirsch M, Lynch C, Mathews W, Manivel JC, Mantyh PW, Clohisy DR. Radiation treatment decreases bone cancer pain through direct effect on tumor cells. Radiat Res 2005;164:400–8.
124. Dudek AZ, Zwolak P, Jasinski P, et al. Protein kinase C-beta inhibitor enzastaurin (LY317615.HCI) enhances radiation control of murine breast cancer in an orthotopic model of bone metastasis. Invest New Drugs 2008;26:13–24.
125. Zwolak P, Dudek AZ, Bodempudi VD, et al. Local irradiation in combination with bevacizumab enhances radiation control of bone destruction and cancer-induced pain in a model of bone metastases. Int J Cancer 2008;122:681–8.
126. Seong J, Park HC, Kim J, Kim UJ, Lee BW. Radiation-induced alteration of pain-related signals in an animal model with bone invasion from cancer. Ann N Y Acad Sci 2004;1030:179–86.
127. Park HC, Seong J, An JH, Kim J, Kim UJ, Lee BW. Alteration of cancer pain-related signals by radiation: proteomic analysis in an animal model with cancer bone invasion. Int J Radiat Oncol Biol Phys 2005;61:1523–34.
128. Fox A, Medhurst S, Courade JP, et al. Antihyperalgesic activity of the cox-2 inhibitor lumiracoxib in a model of bone cancer pain in the rat. Pain 2004;107:33–40.
129. Beyreuther BK, Callizot N, Brot MD, Feldman R, Bain SC, Stohr T. Antinociceptive efficacy of lacosamide in rat models for tumor- and chemotherapy-induced cancer pain. Eur J Pharmacol 2007;565:98–104.
130. Donovan-Rodriguez T, Urch CE, Dickenson AH. Evidence of a role for descending serotonergic facilitation in a rat model of cancer-induced bone pain. Neurosci Lett 2006;393: 237–42.
131. Donovan-Rodriguez T, Dickenson AH, Urch CE. Gabapentin normalizes spinal neuronal responses that correlate with behavior in a rat model of cancer-induced bone pain. Anesthesiology 2005;102:132–40.
132. Sessle BJ. Glia: non-neural players in orofacial pain. J Orofac Pain 2007;21:169–70.
133. Watkins LR, Hutchinson MR, Ledeboer A, Wieseler-Frank J, Milligan ED, Maier SF. Norman Cousins Lecture. Glia as the "bad guys": implications for improving clinical pain control and the clinical utility of opioids. Brain Behav Immun 2007;21:131–46.
134. Watkins LR, Milligan ED, Maier SF. Glial activation: a driving force for pathological pain. Trends Neurosci 2001;24:450–5.
135. Zhang RX, Liu B, Wang L, et al. Spinal glial activation in a new rat model of bone cancer pain produced by prostate cancer cell inoculation of the tibia. Pain 2005;118:125–36.
136. Zhang H, Cang CL, Kawasaki Y, et al. Neurokinin-1 receptor enhances TRPV1 activity in primary sensory neurons via PKCepsilon: a novel pathway for heat hyperalgesia. J Neurosci 2007;27:12067–77.
137. Zhang RX, Li A, Liu B, et al. Electroacupuncture attenuates bone-cancer-induced hyperalgesia and inhibits spinal preprodynorphin expression in a rat model. Eur J Pain 2008;12:870–8.

138. Blouin S, Basle MF, Chappard D. Rat models of bone metastases. Clin Exp Metastasis 2005;22:605–14.
139. Coleman RE. Management of bone metastases. Oncologist 2000;5:463–70.
140. Bauerle T, Adwan H, Kiessling F, Hilbig H, Armbruster FP, Berger MR. Characterization of a rat model with site-specific bone metastasis induced by MDA-MB-231 breast cancer cells and its application to the effects of an antibody against bone sialoprotein. Int J Cancer 2005;115:177–86.
141. Liepe K, Geidel H, Haase M, Hakenberg OW, Runge R, Kotzerke J. New model for the induction of osteoblastic bone metastases in rat. Anticancer Res 2005;25:1067–73.
142. Liepe K, Runge R, Kotzerke J. The benefit of bone-seeking radiopharmaceuticals in the treatment of metastatic bone pain. J Cancer Res Clin Oncol 2005;131:60–6.
143. Jemal A, Murray T, Ward E, et al. Cancer statistics, 2005. CA Cancer J Clin 2005;55: 10–30.
144. Freedman SD, Callery MP. Central control of pancreatic cancer pain: protecting the enemy? Gastroenterology 2007;132:1191–2.
145. Ceyhan GO, Michalski CW, Demir IE, Muller MW, Friess H. Pancreatic pain. Best Pract Res Clin Gastroenterol 2008;22: 31–44.
146. di Mola FF, di Sebastiano P. Pain and pain generation in pancreatic cancer. Langenbecks Arch Surg 2008;393(6):919–22.
147. Hartel M, di Mola FF, Selvaggi F, et al. Vanilloids in pancreatic cancer: potential for chemotherapy and pain management. Gut 2006;55:519–28.
148. Kuraishi Y, Iida Y, Zhang HW, et al. Suppression by gabapentin of pain-related mechano-responses in mice given orthotopic tumor inoculation. Biol Pharm Bull 2003;26:550–2.
149. Andoh T, Sugiyama K, Fujita M, et al. Pharmacological evaluation of morphine and non-opioid analgesic adjuvants in a mouse model of skin cancer pain. Biol Pharm Bull 2008;31:520–2.
150. Martin LA, Hagen NA. Neuropathic pain in cancer patients: mechanisms, syndromes, and clinical controversies. J Pain Symptom Manage 1997;14:99–117.
151. Shimoyama M, Tanaka K, Hasue F, Shimoyama N. A mouse model of neuropathic cancer pain. Pain 2002;99:167–74.
152. Zhu Z, Friess H, diMola FF, et al. Nerve growth factor expression correlates with perineural invasion and pain in human pancreatic cancer. J Clin Oncol 1999;17:2419–28.
153. Honore P. Behavioral assessment of neuropathic pain in preclinical models. Drug Dev Res 2006;67:302–7.
154. Quasthoff S, Hartung HP. Chemotherapy-induced peripheral neuropathy. J Neurol 2002;249:9–17.
155. Aley KO, Reichling DB, Levine JD. Vincristine hyperalgesia in the rat: a model of painful vincristine neuropathy in humans. Neuroscience 1996;73:259–65.
156. Tanner KD, Levine JD, Topp KS. Microtubule disorientation and axonal swelling in unmyelinated sensory axons during vincristine-induced painful neuropathy in rat. J Comp Neurol 1998;395:481–92.
157. Cavaletti G, Bogliun G, Crespi V, et al. Neurotoxicity and ototoxicity of cisplatin plus paclitaxel in comparison to cisplatin plus cyclophosphamide in patients with epithelial ovarian cancer. J Clin Oncol 1997;15: 199–206.
158. Cavaletti G, Tredici G, Braga M, Tazzari S. Experimental peripheral neuropathy induced in adult rats by repeated intraperitoneal administration of taxol. Exp Neurol 1995; 133:64–72.
159. Mimura Y, Kato H, Eguchi K, Ogawa T. Schedule dependency of paclitaxel-induced neuropathy in mice: a morphological study. Neurotoxicology 2000;21:513–20.
160. Strumberg D, Brugge S, Korn MW, et al. Evaluation of long-term toxicity in patients after cisplatin-based chemotherapy for non-seminomatous testicular cancer. Ann Oncol 2002;13:229–36.
161. Polomano RC, Bennett GJ. Chemotherapy-evoked painful peripheral neuropathy. Pain Med 2001;2:8–14.
162. Higuera ES, Luo ZD. A rat pain model of vincristine-induced neuropathy. Methods Mol Med 2004;99:91–8.
163. Authier N, Coudore F, Eschalier A, Fialip J. Pain related behaviour during vincristine-induced neuropathy in rats. Neuroreport 1999;10:965–8.
164. Joseph EK, Levine JD. Sexual dimorphism for protein kinase c epsilon signaling in a rat model of vincristine-induced painful peripheral neuropathy. Neuroscience 2003;119: 831–8.
165. Tanner KD, Reichling DB, Levine JD. Nociceptor hyper-responsiveness during vincristine-induced painful peripheral neuropathy in the rat. J Neurosci 1998;18:6480–91.

166. Topp KS, Tanner KD, Levine JD. Damage to the cytoskeleton of large diameter sensory neurons and myelinated axons in vincristine-induced painful peripheral neuropathy in the rat. J Comp Neurol 2000;424:563–76.
167. Nozaki-Taguchi N, Chaplan SR, Higuera ES, Ajakwe RC, Yaksh TL. Vincristine-induced allodynia in the rat. Pain 2001;93: 69–76.
168. Lynch JJ,3rd, Wade CL, Zhong CM, Mikusa JP, Honore P. Attenuation of mechanical allodynia by clinically utilized drugs in a rat chemotherapy-induced neuropathic pain model. Pain 2004;110:56–63.
169. Kohler DR, Goldspiel BR. Paclitaxel (taxol). Pharmacotherapy 1994;14:3–34.
170. Rowinsky EK, Eisenhauer EA, Chaudhry V, Arbuck SG, Donehower RC. Clinical toxicities encountered with paclitaxel (taxol). Semin Oncol 1993;20:1–15.
171. Rowinsky EK, Chaudhry V, Cornblath DR, Donehower RC. Neurotoxicity of taxol. J Natl Cancer Inst Monogr 1993;15:107–15.
172. Rowinsky EK, Donehower RC. The clinical pharmacology of paclitaxel (taxol). Semin Oncol 1993;20:16–25.
173. Socinski MA, Langer CJ. Single-agent paclitaxel and paclitaxel/non-platinum combination therapy in advanced non-small cell lung cancer. Semin Oncol 1999;26:51–61; discussion 71–2.
174. Wiernik PH, Schwartz EL, Einzig A, Strauman JJ, Lipton RB, Dutcher JP. Phase I trial of taxol given as a 24-hour infusion every 21 days: responses observed in metastatic melanoma. J Clin Oncol 1987;5: 1232–9.
175. Sahenk Z, Barohn R, New P, Mendell JR. Taxol neuropathy. Electrodiagnostic and sural nerve biopsy findings. Arch Neurol 1994;51:726–9.
176. Polomano RC, Mannes AJ, Clark US, Bennett GJ. A painful peripheral neuropathy in the rat produced by the chemotherapeutic drug, paclitaxel. Pain 2001;94:293–304.
177. Ueda H. Molecular mechanisms of neuropathic pain-phenotypic switch and initiation mechanisms. Pharmacol Ther 2006;109: 57–77.
178. Tamura T, Sasaki Y, Nishiwaki Y, Saijo N. Phase I study of paclitaxel by three-hour infusion: hypotension just after infusion is one of the major dose-limiting toxicities. Jpn J Cancer Res 1995;86:1203–9.
179. Ohtsu T, Sasaki Y, Tamura T, et al. Clinical pharmacokinetics and pharmacodynamics of paclitaxel: a 3-hour infusion versus a 24-hour infusion. Clin Cancer Res 1995;1:599–606.
180. Rowinsky EK, Donehower RC. Paclitaxel (taxol). N Engl J Med 1995;332:1004–14.
181. Wiernik PH, Schwartz EL, Strauman JJ, Dutcher JP, Lipton RB, Paietta E. Phase I clinical and pharmacokinetic study of taxol. Cancer Res 1987;47:2486–93.
182. De Brabander M, Geuens G, Nuydens R, Willebrords R, De Mey J. Taxol induces the assembly of free microtubules in living cells and blocks the organizing capacity of the centrosomes and kinetochores. Proc Natl Acad Sci U S A 1981;78:5608–12.
183. Kumar N. Taxol-induced polymerization of purified tubulin. Mechanism of action. J Biol Chem 1981;256:10435–41.
184. Horwitz SB. Mechanism of action of taxol. Trends Pharmacol Sci 1992;13:134–6.
185. Boyle FM, Wheeler HR, Shenfield GM. Amelioration of experimental cisplatin and paclitaxel neuropathy with glutamate. J Neurooncol 1999;41:107–16.
186. Cliffer KD, Siuciak JA, Carson SR, et al. Physiological characterization of Taxol-induced large-fiber sensory neuropathy in the rat. Ann Neurol 1998;43:46–55.
187. Matsumoto M, Inoue M, Hald A, Xie W, Ueda H. Inhibition of paclitaxel-induced A-fiber hypersensitization by gabapentin. J Pharmacol Exp Ther 2006;318:735–40.
188. Smith SB, Crager SE, Mogil JS. Paclitaxel-induced neuropathic hypersensitivity in mice: responses in 10 inbred mouse strains. Life Sci 2004;74:2593–604.
189. Apfel SC, Lipton RB, Arezzo JC, Kessler JA. Nerve growth factor prevents toxic neuropathy in mice. Ann Neurol 1991;29:87–90.
190. Authier N, Fialip J, Eschalier A, Coudore F. Assessment of allodynia and hyperalgesia after cisplatin administration to rats. Neurosci Lett 2000;291:73–6.
191. Gispen WH, Hamers FP, Vecht CJ, Jennekens FG, Neyt JP. ACTH/MSH like peptides in the treatment of cisplatin neuropathy. J Steroid Biochem Mol Biol 1992;43:179–83.
192. de Koning P, Neijt JP, Jennekens FG, Gispen WH. Org.2766 protects from cisplatin-induced neurotoxicity in rats. Exp Neurol 1987;97:746–50.
193. Cata JP, Weng HR, Dougherty PM. Behavioral and electrophysiological studies in rats with cisplatin-induced chemoneuropathy. Brain Res 2008;1230:91–8.
194. Fischer SJ, Podratz JL, Windebank AJ. Nerve growth factor rescue of cisplatin neurotoxicity is mediated through the high affinity receptor: studies in PC12 cells and p75 null mouse dorsal root ganglia. Neurosci Lett 2001; 308:1–4.

195. McDonald ES, Windebank AJ. Cisplatin-induced apoptosis of DRG neurons involves bax redistribution and cytochrome c release but not fas receptor signaling. Neurobiol Dis 2002;9:220–33.
196. ter Laak MP, Hamers FP, Kirk CJ, Gispen WH. rhGGF2 protects against cisplatin-induced neuropathy in the rat. J Neurosci Res 2000;60:237–44.
197. Pradat PF, Kennel P, Naimi-Sadaoui S, et al. Viral and non-viral gene therapy partially prevents experimental cisplatin-induced neuropathy. Gene Ther 2002;9:1333–7.
198. Tassler P, Dellon AL, Lesser GJ, Grossman S. Utility of decompressive surgery in the prophylaxis and treatment of cisplatin neuropathy in adult rats. J Reconstr Microsurg 2000;16:457–63.
199. Callizot N, Andriambeloson E, Glass J, et al. Interleukin-6 protects against paclitaxel, cisplatin and vincristine-induced neuropathies without impairing chemotherapeutic activity. Cancer Chemother Pharmacol 2008;62: 995–1007.
200. Karai L, Brown DC, Mannes AJ, et al. Deletion of vanilloid receptor 1-expressing primary afferent neurons for pain control. J Clin Invest 2004;113:1344–52.
201. Khasabov SG, Ghilardi JR, Mantyh PW, Simone DA. Spinal neurons that express NK-1 receptors modulate descending controls that project through the dorsolateral funiculus. J Neurophysiol 2005;93: 998–1006.
202. Allen JW, Mantyh PW, Horais K, et al. Safety evaluation of intrathecal substance P-saporin, a targeted neurotoxin, in dogs. Toxicol Sci 2006;91:286–98.
203. Mouedden ME, Meert TF. Pharmacological evaluation of opioid and non-opioid analgesics in a murine bone cancer model of pain. Pharmacol Biochem Behav 2007;86:458–67.
204. Zha S, Yegnasubramanian V, Nelson WG, Isaacs WB, De Marzo AM. Cyclooxygenases in cancer: progress and perspective. Cancer Lett 2004;215:1–20.
205. Wood LJ, Nail LM, Gilster A, Winters KA, Elsea CR. Cancer chemotherapy-related symptoms: evidence to suggest a role for proinflammatory cytokines. Oncol Nurs Forum 2006;33:535–42.
206. Body JJ, Mancini I. Bisphosphonates for cancer patients: why, how, and when? Support Care Cancer 2002;10:399–407.
207. Fulfaro F, Casuccio A, Ticozzi C, Ripamonti C. The role of bisphosphonates in the treatment of painful metastatic bone disease: a review of phase III trials. Pain 1998;78: 157–69.
208. Clohisy DR, Mantyh PW. Bone cancer pain and the role of RANKL/OPG. J Musculoskelet Neuronal Interact 2004;4:293–300.
209. Blair JM, Zhou H, Seibel MJ, Dunstan CR. Mechanisms of Disease: roles of OPG, RANKL and RANK in the pathophysiology of skeletal metastasis. Nat Clin Pract Oncol 2006;3:E1.
210. Nagae M, Hiraga T, Yoneda T. Acidic microenvironment created by osteoclasts causes bone pain associated with tumor colonization. J Bone Miner Metab 2007;25:99–104.
211. Lacroix-Fralish ML, Ledoux JB, Mogil JS. The Pain Genes Database: an interactive web browser of pain-related transgenic knockout studies. Pain 2007;131:3.e1–3.e4.
212. Kontinen VK, Kalso E. Of mice and men: what can we predict from the effects of morphine in a mouse model of bone cancer? Pain 2007;132:5–7.
213. De Souza EB, Cload ST, Pendergrast PS, Sah DW. Novel therapeutic modalities to address nondrugable protein interaction targets. Neuropsychopharmacology 2009;34: 142–58.

Chapter 9

Animal Models of Diabetic Neuropathic Pain

Maxim Dobretsov, Miroslav (Misha) Backonja, Dmitry Romanovsky, and Joseph R. Stimers

Abstract

Pain is frequently the earliest and most problematic syndrome of distal peripheral neuropathy (DPN) in diabetic patients. The variety of time of onset, duration and progression, modalities, and severity of individual presentations of painful DPN makes classification and evaluation of mechanisms of pain in patients with diabetic neuropathy an outstandingly difficult task. One critical step to address this issue is the need for large-scale prospective studies that start in pain-free prediabetic or diabetic subjects and use questionnaires and neurological bedside and quantitative sensory tests standardized to assess progression of all possible modalities and types of pain. By their nature, however, even the best equipped and designed clinical studies remain mostly observational, and in-depth understanding of human disease is not possible without studies in animal models. Over the past several decades, a wide variety of rodent models of diabetes have been developed and characterized. Progression of DPN in many of these models has also been studied and confirmed, and in this work, we review those data with a specific focus on the utility, challenges, and limitations of using rodent models in research on mechanisms of diabetic pain.

1. Diabetes, Peripheral Neuropathy, and Pain

Diabetes is a group of metabolic diseases characterized by hyperglycemia resulting from defects in insulin secretion, insulin action, or both (type 1, type 2, and advanced stages of both types of the disease, respectively, (1)). Diabetic neuropathy is defined as "the presence of symptoms and/or signs of peripheral nerve dysfunction in people with diabetes after the exclusion of other causes" (2). The most common diabetic neuropathy is distal peripheral neuropathy (DPN, about 70% of all cases (3)); the disease is distinguishable by symmetrical presentation and distal-to-proximal progression of symptoms of sensorimotor impairment. From the combined data on prevalence (Table 1), on average, at least half of patients with DPN have pain as one manifestation of their neuropathy. Furthermore, in about 30% of these cases of painful DPN (PDPN), the pain is graded as severe, in 50% as moderate, and

Chao Ma and Jun-Ming Zhang (eds.), *Animal Models of Pain*, Neuromethods, vol. 49,
DOI 10.1007/978-1-60761-880-5_9, © Springer Science+Business Media, LLC 2011

Table 1
Type of diabetes and average prevalence[a] of clinical DPN

Diabetes	Prevalence of DPN in diabetes Mean ± SE (number of sources) DPN, all cases	 PDPN	Prevalence of pain in DPN (%)
Type 1	22 ± 3% (9)	11 ± 5% (2)	50
Type 2	26 ± 6% (11)	16 ± 5% (4)	62

[a]Average data from (4, 5, 11, 18, 25, 116–126), number of sources used is indicated in brackets

only in 20% as mild (4–6). At the same time, however, in the general community clinic, 35–60% of PDPN patients receive no treatment other than therapy to reduce blood glucose, and the latter unfortunately does not appear to relieve the pain. Moreover, the efficacy of pain-reducing therapies in the remaining cases does not exceed 30–50% of pain reduction (7–11).

Development of an optimal treatment strategy requires knowledge of mechanism and natural history of the disease. Furthermore, since it is not possible to predict whether different symptomatic presentations of the disease have common or distinct mechanisms, the natural history of each clinical symptom and sign should be traced back to its onset independently (12–14). Because of the complexity of background disease (diabetes) and difficulty of detection of preclinical diabetes and neuropathy, no symptoms of PDPN has yet received such characterization.

2. Natural History and Symptoms of Human PDPN

Diabetes is a result of slowly progressing deterioration in control of peripheral functions of insulin. The disease passes through an asymptomatic stage of early prediabetes, followed by a stage of moderate fasting and/or postprandial hyperglycemia (advanced prediabetes) and culminates in a state of overt and chronic hyperglycemia (Fig. 1). Once diagnosed, diabetes is treated to normalize blood glucose. Risk of hypoglycemia, however, does not allow complete correction and in terms of glucose metabolism, controlled diabetes is more similar to advanced prediabetes than either normal or overtly diabetic state.

Onset of pain may occur at any time following diagnosis of diabetes. Furthermore, pain is a frequently presenting manifestation of neuropathy, leading to discovery and diagnosis of DPN and diabetes or advanced prediabetes in otherwise asymptomatic

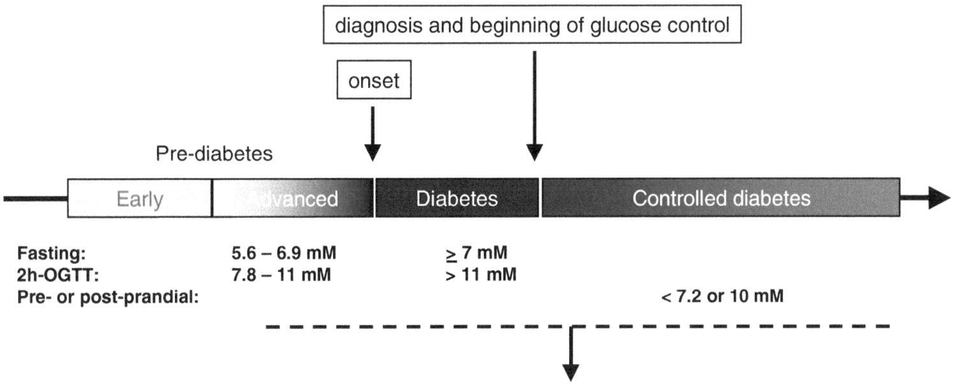

Spontaneous pain:	
Dull, deep aching pain	89-92%
Pain paroxysms	77-81%
Burning pain	23-61%
Evoked pain:	
Pain on pressure	71%
Contact discomfort (sensation loss)	31% (69%)
Pinprick hyperalgesia (hypoalgesia)	23% (34%)
Heat hyperalgesia (hypoalgesia)	0-65% (15%)
Cold hyperalgesia (hypoalgesia)	0-3% (51%)
Warm allodynia (sensation loss)	0% (15-82%)
Cold allodynia (sensation loss)	0% (39-58%)

Fig. 1. Progression of diabetes and symptoms of PDPN in human patients. Fasting and 2 h oral glucose tolerance test (OGTT) as well as pre- and postprandial plasma glucose cutoff values used to diagnose prediabetes and diabetes and as goals of glycemic control are as recommended by American Diabetes Association (ADA (127)). Data on prevalence of painful and related symptoms manifesting as positive or negative phenomena (*in parentheses*) are from (4, 5, 11, 18, 19, 25, 116–118, 128–132).

patients (15). Except for rare cases of acute/remitting painful neuropathy (16, 17), pain symptoms once occurred last for years, demonstrating no significant changes in either severity or modality. Furthermore, no detectable relation exists between pain symptoms and either duration of diabetes or degree of glycemic impairment or severity of other symptoms of DPN (chronic PDPN (8, 11, 18–24)). What triggers pain and whether the same pain-precipitating conditions operate in patients with long-standing diabetes and in patients with advanced prediabetes or new-onset diabetes is not known at this time.

Further difficulty is added by a symptomatic variability of individual cases of PDPN. Prevalence of major manifestations of pain in cases of chronic PDPN is shown in Fig. 1. Deep aching pain and pain on pressure appear as the most common complaints within categories of spontaneous and evoked pains, respectively. However, isolated modalities of pain occur rarely with 95% of PDPN patients experiencing two or all three categories of spontaneous pains (25, 26), and available data on possible clustering of different symptoms of DPN are sketchy at best (Fig. 2).

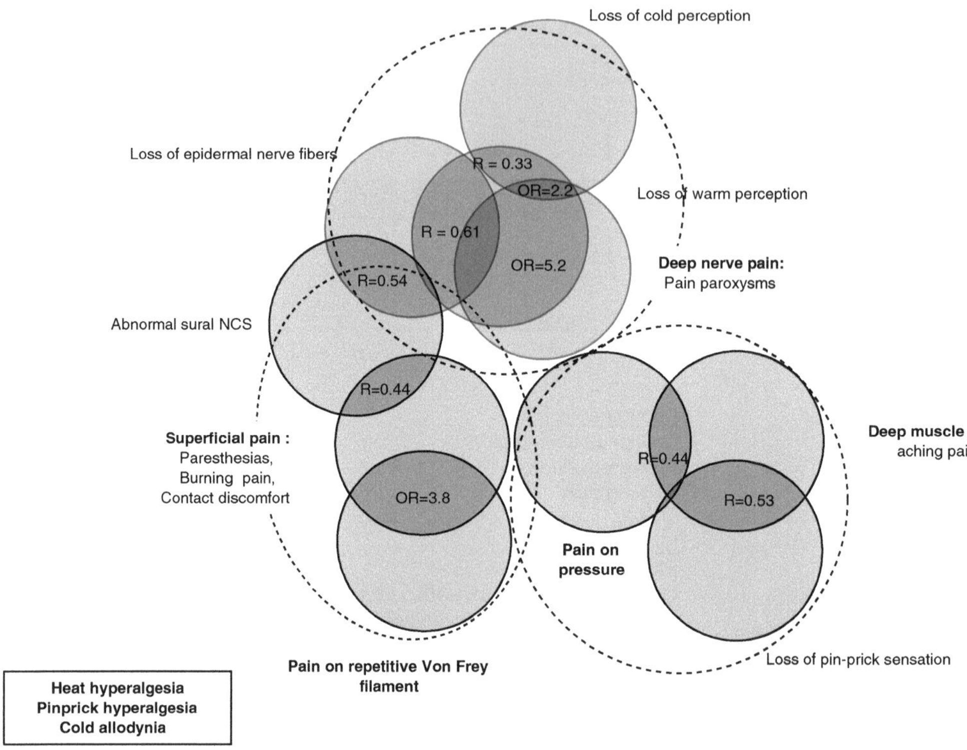

Fig. 2. Clustering of signs and symptoms in PDPN. Available data on regression coefficients and odds ratios (19, 26, 28, 51) suggest existence of three clusters of spontaneous and evoked pain symptoms and negative signs of DPN. The *shaded circle radius* is an approximate equivalent of regression coefficient of 0.35 or odds ratio of 1.75, and *dashed circles* represent potential clusters (still to be mathematically proven).

Nonetheless, even these limited data raise many important questions and clearly demonstrate the need for evaluation of individual pain symptoms. Thus, it appears that aching pain, pain paroxysms, and burning pain/tactile allodynia may indeed represent culmination of different pathologic mechanisms and require different treatment approaches as it has been suggested previously (26). It is also very tempting to conclude that heat and pinprick hyperalgesia and cold allodynia are early and probably transient pain phenomena that are relatively rare in chronic PDPN and show no association with spontaneous pain (see also (27)).

Clinical evaluation of mechanisms of pain faces multiple challenges. Pain is an individual experience, and patients tend to report most severe and chronic conditions at the expense of mild or short-lived symptoms (26, 28). Use of combined symptom scores convenient for the diagnosis and staging of DPN but completely lacking information on prevalence of specific signs and manifestations of the disease adds another problem. There is also general preoccupation with clinical and laboratory assessment of negative symptoms and signs of DPN with little attention paid to evoked pain phenomena (29). Yet another challenge is that the amount of available clinical information is intrinsically proportional to the prevalence of human cases of the disease. As a result, most of what we know about PDPN comes from studies of chronic PDPN in type 2 diabetes. While well-designed prospective studies are needed for understanding diabetic pain in general, it is specifically true with respect to evaluation of PDPN associated with type 1 diabetes and prediabetic stages of the disease of both types (30). The lack of a thorough clinical characterization of diabetic pain symptoms goes beyond just being a clinical problem. Development and evaluation of predictive validity (the ability to predict the human phenomenon) of animal models, "requires parallel development of clinical measures that allow meaningful comparisons" (31). From the discussion below, we see that there are unfortunately too many situations when the potential usefulness of the animal model cannot be estimated because of insufficiency of relevant human data.

3. Diabetes and Painful Neuropathy in Rodent Models

3.1. Rodent Models of Diabetes

Over the decades, a large variety of relatively inexpensive and simple rodent models of diabetes have been developed and characterized, raising the possibility that some difficult-to-address questions of diabetic pain in humans may be more easily resolved in experiments in diabetic animals. Of course, none of the rodent models mimics entirely the human disease; however, Fig. 3 demonstrates that almost all stages of both major types of human diabetes can be reproduced in rodent models with respect to manifestation of at least one pain symptom.

Among STZ-rat, NOD-mice, and BBDP-rat models of diabetes, only BBDP animals develop severe ketoacidosis, requiring insulin therapy (32), and therefore, routinely only BBDP rat is studied under conditions mimicking the stage of controlled diabetes in human patients. However, there is no reason why the stage of controlled diabetes could not be reproduced in any rodent model of overt diabetes. The stage of type 1 advanced prediabetes appears as the least accessible for experimental research. It is relatively short in humans (33), and it lasts for 1 week or less in most rodent models of spontaneous diabetes characterized in this respect. STZ-NG animals rarely progress to advanced prediabetes (34). Although, it remains possible that titrated STZ treatment can be used to induce long-lasting postprandial hyperglycemia in rodents (35).

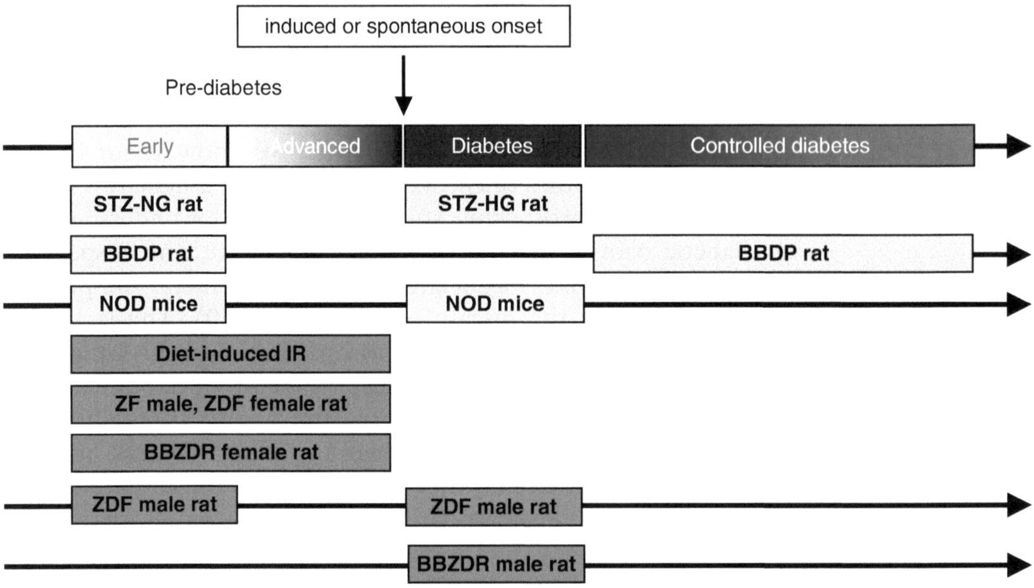

Fig. 3. Stages of diabetes and rodent models. Corresponding stages of human diabetes and diabetes in rodent models are aligned. *Lighter and darker shading* indicates type 1 and 2 disease models, respectively. STZ-NG rat and STZ-HG rat/mouse are models of streptozotocin-induced early prediabetes (normoglycemia; NG (75)) and overt diabetes (hyperglycemia, HG (43, 133)). BBDP rat is biobreeding diabetes-prone Worchester model with spontaneous onset type 1 disease at about tenth week of age (134–138). NOD (nonobese diabetic) mouse is a mouse model of spontaneous type 1 disease developing between 12 and 20 weeks of age similar to BBDP rat (43, 139). Rats with diet (high-calorie food or sucrose-enriched drinking water)-induced insulin resistance (IR (86, 140–147), Zucker fatty (ZF) and young diabetic fatty (ZDF, <7 weeks of age (148–151)) rats are animals suitable in studies of type 2 early and advanced prediabetes. After seventh week of age male ZDF rats and after tenth week of age BBZDR (biobreeding Zucker diabetic) rats spontaneously develop type 2 diabetes (112, 152). Both strains feature hyperinsulinemia, obesity, and dyslipidemia, but hypertension is more advanced in BBZDR than in ZDF rats (58, 112, 152–154).

3.2. Studies of Pain in Rodents (General Problems)

Overall, it is not a shortage of rodent models of diabetes but intrinsic limitations of these models and imbalance of our knowledge of human and rodent PDPN that constitutes the major obstacle in studies of diabetic pain mechanisms in animals.

First, there are easily recognizable changes in animal posture, gait, and behavior (licking, guarding, and autotomy) that manifest unilateral spontaneous pain in rodents with experimental mononeuropathy (36, 37). The diffuse bilateral spontaneous pain is much more difficult to assess (38). Attempts to use measurements of the animal's sleep and grooming patterns, exploratory and social behavior, or ultrasonic vocalization (39–41) have been unreliable indicators of pain in rodents and specifically in type 1 diabetic animals in which behavior is deeply affected by general sickness (42). Thus, studies of pain in animals are limited to measurements of characteristics of withdrawal response and by the assumption that withdrawal reflex is an accurate measure of painfulness of the stimulus presented (32, 38, 43, 44). At the same time, except for rare reports, spontaneous but not evoked pain is what we know most about human PDPN (see previous section).

Table 2
Evoked-pain tests in animal models of diabetes

Evoked pain	Test in animals	Measured characteristic of withdrawal reaction	Animal models studied with test
Pain on pressure	Randall-Selitto pressure algometer (rat only)	Withdrawal threshold, *g*	STZ-rat, Zucker rat, sand rat
Contact discomfort	Static allodynia, graded or electronic Von Frey (Semmes–Weinstein) nylon filament test	Frequency of response, % or withdrawal threshold, *g*	STZ-rat and mice, NOD mice, diet-induced IR (rat)
	Dynamic allodynia, camel brush test	Frequency of response, %	No data
Pin-prick hyperalgesia	Sharp Von Frey filament test or pinprick test	Frequency of response, % or withdrawal threshold, *g*	STZ-rat and mice
Heat hyperalgesia	Hot plate test, water immersion test, or plantar radiant heat test	Withdrawal threshold, °C or latency, *s*	STZ-rat, NOD mice, BBDP rat, Zucker rat, BBZDR rat
Cold allodynia/ hyperalgesia	Cold plate test or acetone drop test	Frequency of response, % or withdrawal latency, *s*	STZ-rat, diet-induced IR (rat)

Furthermore, while methods allowing assessment of nearly all modalities of evoked pain in animals exist (Table 2), not all of these tests are applicable in all cases, and since there is no standard protocol (38), there are no two models of diabetes that have been characterized at similar stages and duration of the disease, using a complete and similar battery of tests and measures of withdrawal reaction. Although undoubtedly attractive because of ease of genetic manipulations (45), use of mouse models in behavioral studies of pain suffers from an additional problem. Unlike rat, the majority of mouse strains become increasingly stressed and agitated with repeated handling, which raises enormous difficulties with interpretation of results obtained in longitudinal studies of evoked pain in these animals (46).

Finally, diabetic STZ rat appears by far as the best studied model with regard to PDPN (43). This creates an additional imbalance between animal and clinical studies focused mostly on type 2 diabetes cases.

Viewed as yet another limitation, the rodent life span is too short to allow rodent models to mimic human DPN that has a slow insidious onset and progresses over decades of human age to culminate in irreversible symptoms and multiple morphological signs of nerve injury. Functional nerve changes observed in diabetic rodents have too abrupt onset; they are too easy to reverse by treatments not working in humans, and their profile seems to be too uniform to be compatible to human conditions (44, 47).

Considering these critiques, it should be noted that the situation may turn out to be not that deplorable if the time scale of progression of diabetes in rodent models is adjusted in accord with differences in human and rodent life spans and diverse genetic background in the human population vs. uniform in studied rodent strains are taken into account. As an example of the former, Fig. 4 uses two horizontal scales, lower scale showing duration of diabetes in rats, and upper scale showing human equivalent, calculated assuming 2.5 and 75 years as average duration of rat and human life spans. To what extent such scaling is justified requires further research (38). However, accumulation of reactive oxygen species (ROS) is considered to be one of the major determinants of both aging and diabetic neuropathy

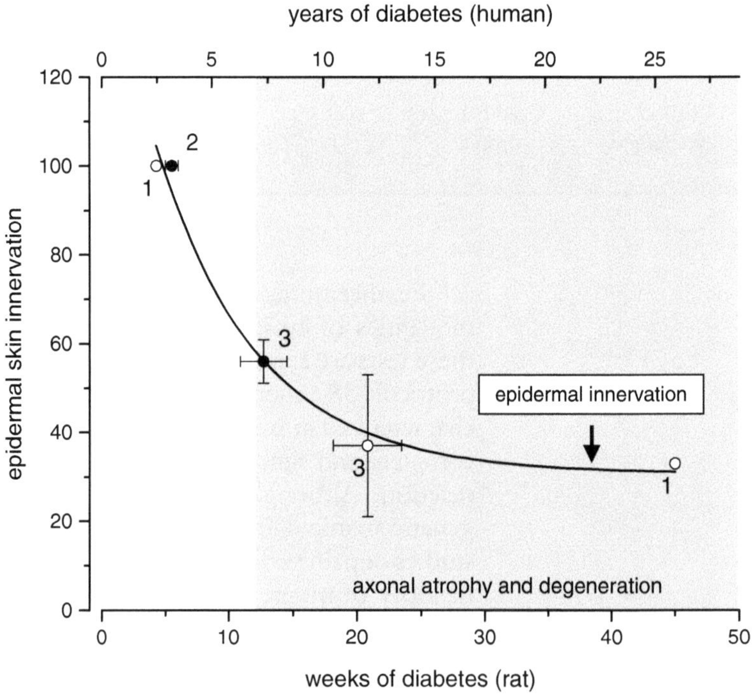

Fig. 4. Diabetes and neurodegeneration in rats and humans. Line is a best fit of exponential curve to the data on epidermal density of nerve fibers (EDNF; normalized to control) in diabetic humans (*open circles* (54–57)) and STZ rats (*closed circles* (51, 52, 56, 57)); symbols with error bars represent average of data from several sources (number next to the symbol). Human and rat diabetes duration scales are adjusted in accord with between-species difference in average life span (rat: 2.5 and human: 75 years). First detectable nerve morphological abnormalities are reported at 12th week of diabetes in STZ rat (*left border* of the *grayed out area* in the figure (155, 156)) and on 16th week of diabetes in BBDP rat (157, 158) model.

(48–50). Metabolic rate and associated accumulation of ROS are faster in small as compared to large mammals which can likely explain a shorter life span of the former (49, 50). To the degree that the hypothesis above is correct, we may expect that the consequences of metabolic insults raising the production of ROS (diabetes) will develop faster and manifest within a shorter time in rodents than in humans. Interestingly, in support of this view, with the use of adjusted scales, the progression of epidermal denervation in diabetic humans and STZ rats (filled squares and opened triangles; data from (19, 51–53) and (54–57), respectively) appears to have similar time courses starting at 6 weeks and declining to about 30% of normal by 20 weeks of diabetes (equivalent duration of human disease is 3 and 10 years; Fig. 4).

In addition, although it might be just a remarkable coincidence, impairment of nerve conduction in both rats and humans becomes difficult to reverse after about the same duration of diabetes (8 weeks in rats or 5 years in humans; (58, 59)), and this is about the time when in both species degenerative changes take the stage, Fig. 4 (grayed area). Whether the speculations above are valid, specifically for signs and symptoms of PDPN, is not known.

With regard to profiles of diabetic neuropathy, which is variable in humans and uniform in rodent models, this difference is absolutely superficial, created by the inevitable focusing of research on usually male rodents coming from one or few selected inbred or outbred strains of animals. Multiple studies have demonstrated that inclusion of animals with variable genetic background will immediately increase variation in outcome measures, including susceptibility to obesity-inducing diet and diabetes (60, 61), background sensitivity to painful stimuli, and development of pain phenomena in response to neuropathic insults (46, 62, 63). Linked to genetic analysis, evaluation of specific pain symptoms under the same protocol, but in different strains and sexes of animals and models of diabetes, would be of great value for understanding the mechanisms of human PDPN.

3.3. Studies of Evoked Pain in Rodents

3.3.1. Pain on Pressure

In healthy, adult humans, deep muscle pain on pressure is usually described as cramping; it is well localized and is mediated by activation of group III and IV muscle (Aδ and C in nomenclature for skin afferents) slowly adapting nociceptors (64–66). In rats, a Randall–Selitto algometer is used to measure hind-limb pressure pain withdrawal threshold (PPT, (67)), and a decrease in PPT is interpreted as an indication of pressure hyperalgesia. The method cannot be applied to mice because of small size of the animal and because a tested limb should not be restrained in any way and this could be achieved only by using a repetitive handling and establishing a "trustful bond" between the animal and the investigator. As it has been noted earlier, mice do not tolerate repetitive handling (46).

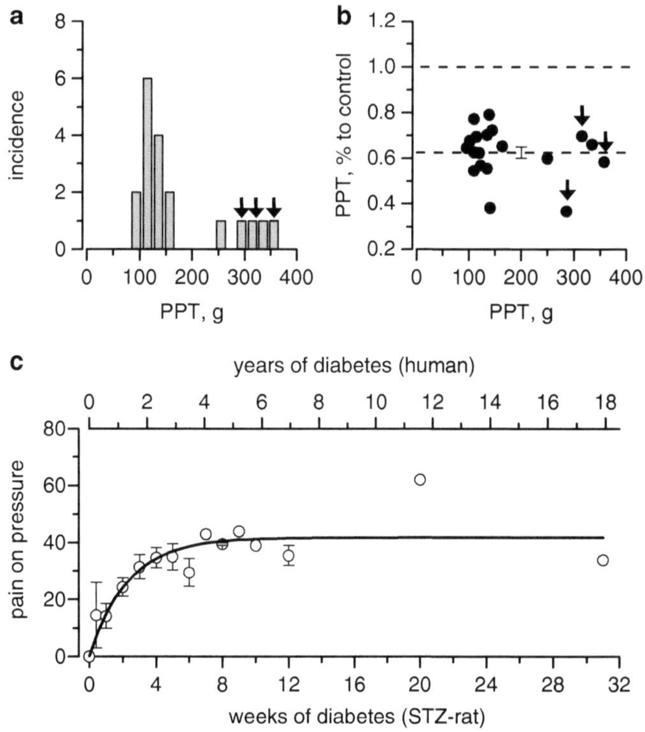

Fig. 5. Pressure pain thresholds in control and STZ diabetic rats. (**a**) Frequency distribution of data from literature on PPT in control rats (total 21 independent publications). *Arrows* mark the data coming from studies explicitly stating vocalization as the test end-point measure (159–161). In the remaining reports, paw withdrawal, vocalization, or overt struggle are used alone or in combination as end-point descriptors. From our experience, PPT of healthy Sprague–Dawley rats rarely exceeds 150 g limit when withdrawal, but not vocalization is the end point and when animal is free to withdraw the paw at any time during the test. (**b**) Pressure hyperalgesia in STZ diabetic rats measured in different reports by different laboratories shows a consistent percent reduction regardless of control PPT levels (PPT is decreased to 63 ± 3% from control after 7 ± 2 weeks of STZ-induced diabetes; average for all plotted data, *dashed line* with error bar). (**c**) Progression of pain on pressure with time (symbols with error bars are average data from two to six sources). Figure uses data from (34, 42, 56, 68, 75–78, 92, 93, 98, 99, 159–167).

PPT measurements are strongly affected by the selection of end-point measurement (limb withdrawal vs. vocalization threshold), and from our experience, they do not tolerate even the slightest restraint of the tested hind limb (Fig. 5a). Interestingly, however, despite the great variation in the literature with regard to control and baseline PPT values, as of now, the pressure hyperalgesia is the most consistent finding in studies in rat type 1 (STZ; Fig. 5b) and type 2 (Sand rat (*Psammomys obesus*) and ZDF, (68, 69)) rat models of diabetes. Furthermore, with reservation that pain on pressure is poorly characterized in humans with diabetes, this finding in rats still appears as potentially relevant to human disease. Pain on pressure (e.g., pain under feet while walking) is reported by 71% of patients with diabetes and PDPN (28), and fibromyalgic pain has 16–19% prevalence in general population of people with either type of diabetes (2% prevalence in healthy control people (70)). Pain on pressure, deep muscle pain, and loss of pinprick sensation forms a distinct cluster among symptoms and signs of diabetic DPN, suggesting that they may have a unique trigger and/or pathogenesis (Fig. 2). Thus far, studies in rats support this suggestion, showing that a decline in PPT starts in prediabetic animals long before and with no association with type of disease or development of either postprandial or overt hyperglycemia in rats (34, 71). One implication of these data is that some nonglycemic factors, for example, insulinopenia and/or insulin resistance, may affect the function of muscle nociceptors either directly or by modifying local intramuscular conditions (71). Rat models appear to be a convenient tool to address this question.

3.3.2. Tactile Allodynia

Mechanisms of contact discomfort, which is frequently reported as burning pain on normally pain-free light touch or brushing the skin (static and dynamic allodynia, respectively), are not known. It is believed that sensitization of spinal cord wide-diapason nociceptive neurons makes them sensitive to collateral input from skin receptors normally not involved with nociception (large myelinated Aβ afferents (72–74)). Clustering of superficial pain and pain on repetitive application of Von Frey filament (Fig. 2, (28)) seems to support state of central sensitization in people with PDPN. In turn, reports of rapid decline of withdrawal thresholds to stimulation with graded Von Frey hairs or with electronic Von Frey apparatus in STZ diabetic rats suggest that this model may be useful in studies of mechanisms of this central sensitization (Fig. 6; (75–83)).

On average, these studies indicate about 60% decrease of VFT after 6.3 ± 0.9 week of diabetes. A similarly strong decrease in VFT was also observed in several studies in a STZ-mouse model of type 1 diabetes (45, 84, 85) and also in studies in the rat model of diet-induced insulin resistance (12 weeks of 10% sucrose in drinking water (86). Thus, exaggerated sensitivity to touch in diabetes is likely not a rat- or a model-specific phenomenon. Furthermore, the latter data from insulin-resistant rats imply that similar to pain on pressure, pathogenesis of tactile allodynia may start in prediabetes prior to onset of overt hyperglycemia.

However, despite apparent consistency, the observations above raise several concerns. First, and perhaps most important, is that in humans with diabetes, the prevalence of contact discomfort is substantially less (30%) than expected from cited animal studies. At the same time, while the majority of patients with PDPN manifest loss of tactile and pinprick sensations (Fig. 1), we found only one report suggesting that sensitivity to Von Frey filaments declines in mice starting fourth to fifth week of STZ-induced diabetes (87). Whether this discord associates with poor knowledge of progression of human PDPN, or with specifics of the disease in rodents, or with the "bias toward neglecting negative

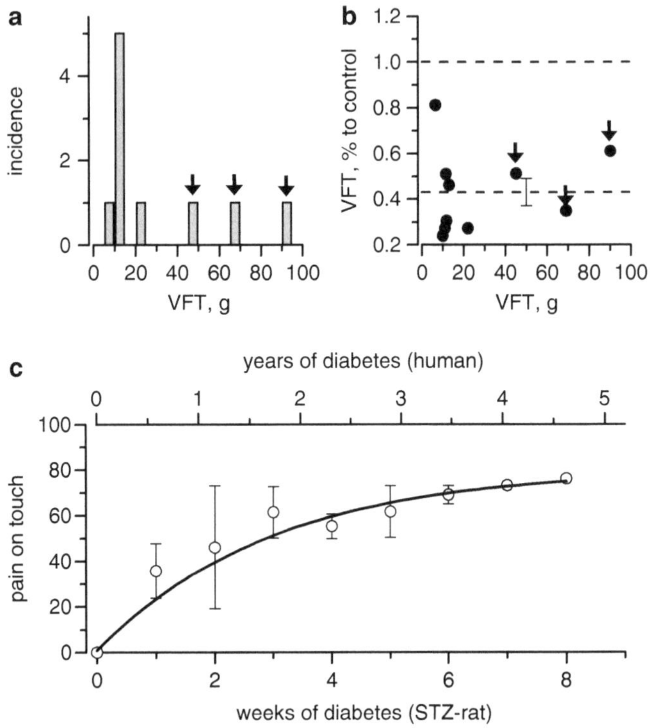

Fig. 6. Pain on light touch (Von Frey threshold) in control and STZ diabetic rats. (**a**) Frequency distributions of tactile withdrawal thresholds measured in control rats using calibrated von Frey filaments or electronic Von Frey technique (*arrows*). (**b**) Both techniques demonstrate a decline in tactile threshold in rats with STZ-induced diabetes. Average for all plotted data is 0.43 ± 0.06 (6.3 ± 0.9 weeks of diabetes). (**c**) Pain on light touch develops within first 2 months of STZ diabetes (symbols with error bars are average data from two to five sources). Figure uses data from (75–83).

studies in the field of experimental science" (88), remains to be seen. Another concern is that with the routinely used test with flat-tip Von Frey filaments of different bending force and also different thickness and sharpness, it is not easy to interpret the event of limb withdrawal as manifestation of allodynia or pure nociceptive flexor response. Mechanical threshold of activation of normal rat skin C- and Aδ-nociceptors are 11 ± 2*g* and 29 ± 2*g*, respectively. Furthermore, STZ-induced diabetes decreases threshold of Aδ nociceptors down to 8 ± 4*g* (89). These values are very close to average from the literature VFT measured in control and diabetic rats (12 and 4.6*g*, see above). Thus, nociceptive component of withdrawal reaction is certainly present during filament testing in normal rats, and amplification of this component (hyperalgesia) may play a substantial role in VFT decrease in diabetic rats. The observation of a bilateral decrease in withdrawal threshold to sharp (100 μm-tip, selective activation of skin C-nociceptors) Von Frey filaments in STZ diabetic and prediabetic rats supports the concern regarding the impact of mechanical hyperalgesia (75).

Finally, this test is extremely sensitive to a great variety of difficult-to-standardize details of experiment (such as duration and rate of application of a filament, site of application on ventral paw, interval between sequential applications, etc.), and therefore, it is exceptionally prone to experimental bias (discussed in (90)). Considering this and also the fact that contact discomfort is frequently a combination of static and dynamic allodynia, more studies comparing withdrawal reactions to a regular Von Frey, sharp Von Frey, and a camel brush stroke (dynamic allodynia) in rat models of diabetes are warranted.

3.3.3. Heat Hyperalgesia

Heat and warm perception reflects the function of small, unmyelinated skin afferents. Loss of this function clusters with pain paroxysms and correlates with total pain rating in humans with diabetes (Fig. 2 and (91)). Heat hyperalgesia has not been linked to any modality of chronic pain, but its prevalence in patients with mild diabetic neuropathy may be as high as 60%, making it second in prevalence after pressure-evoked pain phenomenon (20).

In diabetic rodent models, withdrawal response to noxious heat is one of the most studied and yet controversial issue. In studies of STZ diabetic rat, latency or threshold of hind limb or tail withdrawal to noxious heat were reported to be decreasing (hyperalgesia; (77, 83, 92–97)), not changing (76), temporarily or stably increasing (hypoalgesia; (34, 42, 54, 56, 98–103)) or decreasing during first weeks of diabetes and then increasing (81). In BBDP rat, starting during first month of diabetes and lasting for at least 10 months heat hyperalgesia was observed (104, 105). In the model of diet-induced insulin resistance, no changes in heat paw withdrawal latency was detected after 12 weeks of 10% glucose in drinking water (86), but in prediabetic female Zucker rats tail-flick latency on heat was decreased (106). In type 2 diabetic ZDF rats, plantar radiant heat test detected heat hypoalgesia after 20 weeks of disease (58), while by the same test BBZDR rats demonstrated progressive hyperalgesia starting at 8 weeks and lasting through 32 weeks of type 2 diabetes (105). Experiments in mouse models suffer from similar scatter of the results that are difficult to accommodate within any simple hypothesis (85, 107–111). The only common impression is that prevalence of "negative" (hypoalgesia) observations increases with duration of diabetes, coinciding with onset of decline in epidermal nerve fiber density (ENFD, Fig. 7c). Thus, heat hyperalgesia may indeed be a transient state signaling functional injury of yet morphologically intact skin C fibers (81, 108). This would be in agreement with clinical data showing increased prevalence of heat hyperalgesia in people with moderate but not severe DPN (27). The results of a recent study of STZ-induced diabetes in wild-type and transient receptor potential vanilloid I (TRPVI) – knockout mouse suggests overexpression of TRPVI in skin and nervous system as a reason

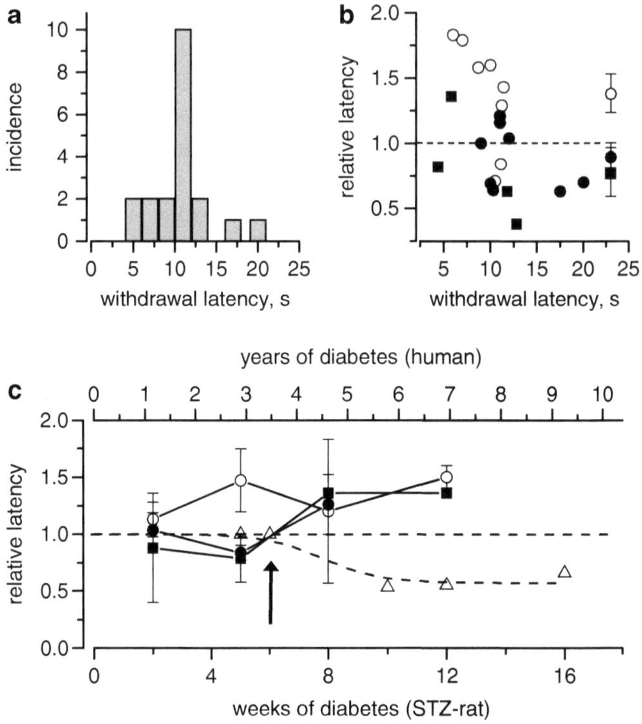

Fig. 7. Heat sensitivity (latency of withdrawal) in control and STZ diabetic rats. (**a**) Frequency distributions of latencies of paw or tail withdrawal measured in hot plate or radiant heat tests in control rats. (**b**) STZ-induced diabetes is reported to produce no changes in heat withdrawal latencies, or decrease or increase them (heat hyperalgesia and hypoalgesia, respectively). The prevalence of observations of hyperalgesic responses to heat in 4–12 week-long experiments is 25, 50, and 75% in studies using hot plate test, or measurements of paw or tail withdrawal in focused beam light tests (*open circles*, or *closed circles* and *squares*; symbols with standard errors on the right side of the panel are average latencies for respective groups of plotted data). (**c**) When the data from different sources are pooled together according to the test used, averaged and plotted vs. duration of STZ-induced diabetes, radiant heat tests (but not hot plate test) suggest the possibility of existence of a transient period of hyperalgesia developing in some animals at about fifth week of diabetes and replaced later with progressive loss of sensitivity to heat. Interestingly, onset of hypoalgesia in these experiments coincides with the onset of epidermal denervation reported by others (*opened triangles*, *dashed line,* and *arrow*; other symbols are as in panel B). Data on heat withdrawal latencies are from (42, 54, 56, 76, 77, 81, 83, 92–102), and data on ENDF are from (51, 52, 56, 57) (symbols with error bars are averages of data from two to eight of these sources).

for transient heat hyperalgesia (108). It would be extremely important to confirm this observation in BBDP rat and NOD mouse in which heat hyperalgesia appears as a stable state lasting for weeks (104, 105, 111) and in type 2 diabetes rat and mouse models in which moderate (105) or no decrease in heat withdrawal latency was observed (58, 107, 112). Another important area to explore is evaluation of heat perception in type 1 and type 2 prediabetic rodents. In NOD mice and BBDP rats, an exaggerated sensitivity to heat seems to predict progression to overt diabetes by at least a few weeks before the onset of disease (95, 111). Similarly, heat hyperalgesia was observed in experiments in female, obese ZDF rats, a model of type 2 prediabetes (106).

As a final comment to this section, the behavioral response to heat is complex. Thus, reactions to fast and slow skin heating are mediated by different, Aδ and C fibers, respectively (113). The tail withdrawal, simple withdrawal of paw (focused light beam test), and withdrawal and licking the paw (hot-plate test) are behaviors with very different levels of supraspinal involvement and therefore potentially very different responses to a disease state (114). Although the analysis of the literature data does not allow attribution of scatter of the measured heat latencies to any single aspect of experimental protocols, the lack of standardized protocols (use of different tests: paw or tail withdrawal, water immersion, hot plate, focused radiant heat tests with different and frequently not specified settings, such as baseline temperature or rate of heat ramp in focused radiant heat test) undoubtedly contributes to the overall confusing picture of nociceptive heat perception in diabetic rodents.

3.3.4. Other Evoked Pain Modalities

Peripheral neuropathy may manifest with dynamic allodynia (as already mentioned) and abnormal sensitivity to cold (30, 115). Although techniques assessing the presence of these abnormalities in rodents exist, they have never been routinely applied in experiments in diabetic rodents. We are aware of just one report addressing the issue of dynamic allodynia. In this work, no abnormalities in frequency of withdrawal to stroking the paw dorsum with a cotton-tipped swab was detected in NOD mouse after 21 weeks of diabetes. In the same work, no signs of cold allodynia assessed by ether-drop application technique were found (109). Cold allodynia, however, was detected by 10°C water tail-immersion test in rats with STZ-induced diabetes (92) and by acetone drop test in rats with diet-induced (sucrose in drinking water) prediabetes (86). Clearly, further experiments are needed to substantiate these findings.

4. Conclusion

Animal models are necessary to determine triggers and dissect cellular and molecular mechanisms of pathogenesis of human disease. This knowledge could then be used to suggest targeted preventive and treatment strategies and to test these strategies in preclinical trials. With respect to diabetic pain, multiple relatively simple rodent models of diabetes are available that are not expensive and mimic certain aspects of evoked pain symptoms

Table 3
Disaccord of studies of diabetic pain in humans and rodents

Categories	Best studied: In humans	In rodents
Type of diabetes	T2DM	T1DM (STZ-induced)
Stage of diabetes	Controlled	Overt
Duration of diabetes	Chronic	Acute
Symptoms of pain	Spontaneous	Evoked

associated with human disease. Furthermore, despite the inherent limitations to studies in rodents (short duration of the disease and treatment, lack of reliable tests for spontaneous pain, etc.), the analysis and comparison of results of experimental studies and clinical trials of treatments directed against evoked pain manifestations has demonstrated surprisingly good predictive validity of STZ rat model of diabetes (88). In addition, these models appear to have the potential to address the question about very first steps of human, prediabetic PDPN (71). Prediabetes and specifically early prediabetes period of disease is difficult for clinical studies, because of the lack of tools of detection of this essentially asymptomatic stage of diabetes in the general population. There is, however, a major obstacle to overcome before the verdict on clinical relevance of findings in rat and mouse models of diabetes can be achieved – this is an imbalance of our knowledge of human and experimental diabetic PDPN (Table 3). Both clinical and basic science efforts will be needed to address this issue. Studies of pain in rodents with controlled hyperglycemia and with an additional emphasis on type 2 diabetes models will help to narrow the gap in the first two categories in Table 3. Nonetheless, only well-designed clinical studies of PDPN in patients with newly diagnosed diabetes and with a balanced assessment of all modalities of spontaneous and evoked pain may help to address the other two concerns. In addition, both clinical and animal studies of diabetic PDPN are obviously in need of standardization of protocols, tests, and end points (with an equal attention devoted to positive and negative, or gain and loss-of-function signs of the disease; see also (38)). Finally, there is a need for further emphasis on studies of prediabetic stage of disease that remains a gray area of our knowledge.

Acknowledgements

This work was supported by NIH NIDDK (grant DK067284) and by UAMS COM funds.

References

1. American Diabetes Association. Diagnosis and classification of diabetes mellitus. Diabetes Care 2008; 31(1 Suppl):S55–S78.
2. Boulton AJ, Malik RA, Arezzo JC, Sosenko JM. Diabetic somatic neuropathies. Diabetes Care 2004; 27:1458–1483.
3. National Diabetes Data Group. Diabetes in America. 2nd ed. Bethesda MD: NIH Publication, 1995: 95–1468.
4. Wu EQ, Borton J, Said G, Le TK, Monz B, Rosilio M et al. Estimated prevalence of peripheral neuropathy and associated pain in adults with diabetes in France. Curr Med Res Opin 2007; 23(9):2035–2042.
5. Davies M, Brophy S, Williams R, Taylor A. The prevalence, severity, and impact of painful diabetic peripheral neuropathy in type 2 diabetes. Diabetes Care 2006; 29(7):1518–1522.
6. Tolle T, Xu X, Sadosky AB. Painful diabetic neuropathy: a cross-sectional survey of health state impairment and treatment patterns. J Diabetes Complications 2006; 20(1):26–33.
7. Boulton AJ. Management of diabetic peripheral neuropathy. Clin Diabetes 2005; 23:9–15.
8. Daousi C, Benbow SJ, Woodward A, MacFarlane IA. The natural history of chronic painful peripheral neuropathy in a community diabetes population. Diabet Med 2006; 23(9):1021–1024.
9. Belgrade MJ, Cole BE, McCarberg BH, McLean MJ. Diabetic peripheral neuropathic pain: case studies. Mayo Clin Proc 2006; 81(4 Suppl):S26–S32.
10. Argoff CE, Backonja MM, Belgrade MJ, Bennett GJ, Clark MR, Cole BE et al. Consensus guidelines: treatment planning and options. Mayo Clin Proc 2006; 81(4 Suppl):S12–S25.
11. Daousi C, MacFarlane IA, Woodward A, Nurmikko TJ, Bundred PE, Benbow SJ. Chronic painful peripheral neuropathy in an urban community: a controlled comparison of people with and without diabetes. Diabet Med 2004; 21(9):976–982.
12. Woolf CJ, Mannion RJ. Neuropathic pain: aetiology, symptoms, mechanisms, and management. Lancet 1999; 353(9168):1959–1964.
13. Calcutt NA, Backonja MM. Pathogenesis of pain in peripheral diabetic neuropathy. Curr Diab Rep 2007; 7(6):429–434.
14. Sindrup SH, Jensen TS. Efficacy of pharmacological treatments of neuropathic pain: an update and effect related to mechanism of drug action. Pain 1999; 83(3):389–400.
15. Singleton JR, Smith AG, Russell J, Feldman EL. Polyneuropathy with impaired glucose tolerance: implications for diagnosis and therapy. Curr Treat Options Neurol 2005; 7(1):33–42.
16. Archer AG, Watkins PJ, Thomas PK, Sharma AK, Payan J. The natural history of acute painful neuropathy in diabetes mellitus. J Neurol Neurosurg Psychiatry 1983; 46(6):491–499.
17. Benn SC, Costigan M, Tate S, Fitzgerald M, Woolf CJ. Developmental expression of the TTX-resistant voltage-gated sodium channels Nav1.8 (SNS) and Nav1.9 (SNS2) in primary sensory neurons. J Neurosci 2001; 21:6077–6085.
18. Sorensen L, Molyneaux L, Yue DK. Insensate versus painful diabetic neuropathy: the effects of height, gender, ethnicity and glycaemic control. Diabetes Res Clin Pract 2002; 57(1):45–51.
19. Shun CT, Chang YC, Wu HP, Hsieh SC, Lin WM, Lin YH et al. Skin denervation in type 2 diabetes: correlations with diabetic duration and functional impairments. Brain 2004; 127(Pt 7):1593–1605.
20. Sorensen L, Molyneaux L, Yue DK. The level of small nerve fiber dysfunction does not predict pain in diabetic neuropathy: a study using quantitative sensory testing. Clin J Pain 2006; 22(3):261–265.
21. Sorensen L, Molyneaux L, Yue DK. The relationship among pain, sensory loss, and small nerve fibers in diabetes. Diabetes Care 2006; 29(4):883–887.
22. Benbow SJ, Chan AW, Bowsher D, MacFarlane IA, Williams G. A prospective study of painful symptoms, small-fibre function and peripheral vascular disease in chronic

painful diabetic neuropathy. Diabet Med 1994; 11(1):17–21.
23. Young RJ, Ewing DJ, Clarke BF. Chronic and remitting painful diabetic polyneuropathy. Correlations with clinical features and subsequent changes in neurophysiology. Diabetes Care 1988; 11(1):34–40.
24. Young RJ, Zhou YQ, Rodriguez E, Prescott RJ, Ewing DJ, Clarke BF. Variable relationship between peripheral somatic and autonomic neuropathy in patients with different syndromes of diabetic polyneuropathy. Diabetes 1986; 35(2):192–197.
25. Dyck PJ, Kratz KM, Karnes JL, Litchy WJ, Klein R, Pach JM et al. The prevalence by staged severity of various types of diabetic neuropathy, retinopathy, and nephropathy in a population-based cohort: the Rochester Diabetic Neuropathy Study. Neurology 1993; 43(4):817–824.
26. Pfeifer MA, Ross DR, Schrage JP, Gelber DA, Schumer MP, Crain GM et al. A highly successful and novel model for treatment of chronic painful diabetic peripheral neuropathy. Diabetes Care 1993; 16:1103–1115.
27. Dyck PJ, Dyck PJ, Larson TS, O'Brien PC, Velosa JA. Patterns of quantitative sensation testing of hypoesthesia and hyperalgesia are predictive of diabetic polyneuropathy: a study of three cohorts. Nerve growth factor study group. Diabetes Care 2000; 23(4):510–517.
28. Otto M, Bak S, Bach FW, Jensen TS, Sindrup SH. Pain phenomena and possible mechanisms in patients with painful polyneuropathy. Pain 2003; 101(1–2):187–192.
29. Proceedings of a consensus development conference on standardized measures in diabetic neuropathy. Clinical measures. Neurology 1992; 42:1823–1825.
30. Smith AG, Singleton JR. Impaired glucose tolerance and neuropathy. Neurologist 2008; 14:23–29.
31. Geyer MA, Markou A. Animal models of psychiatric disorders. In: Bloom FE, Kupfer DJ, editors. Psychopharmacology: The Fourth Generation of Progress. New York: Raven Press, 1995: 787–798.
32. Rees DA, Alcolado JC. Animal models of diabetes mellitus. Diabet Med 2005; 22(4):359–370.
33. Sosenko JM, Palmer JP, Greenbaum CJ, Mahon J, Cowie C, Krischer JP et al. Patterns of metabolic progression to type 1 diabetes in the Diabetes Prevention Trial-Type 1. Diabetes Care 2006; 29(3):643–649.
34. Romanovsky D, Cruz NF, Dienel GA, Dobretsov M. Mechanical hyperalgesia correlates with insulin deficiency in normoglycemic streptozotocin-treated rats. Neurobiol Dis 2006; 24:384–394.
35. Junod A, Lambert AE, Stauffacher W, Renold AE. Diabetogenic action of streptozotocin: relationship of dose to metabolic response. J Clin Invest 1969; 48(11):2129–2139.
36. Wang LX, Wang ZJ. Animal and cellular models of chronic pain. Adv Drug Deliv Rev 2003; 55(8):949–965.
37. Eaton M. Common animal models for spasticity and pain. J Rehabil Res Dev 2003; 40(4 Suppl 1):41–54.
38. Rice AS, Cimino-Brown D, Eisenach JC, Kontinen VK, Lacroix-Fralish ML, Machin I et al. Animal models and the prediction of efficacy in clinical trials of analgesic drugs: a critical appraisal and call for uniform reporting standards. Pain 2008; 139(2):243–247.
39. Vissers K, De Jongh R, Hoffmann V, Heylen R, Crul B, Meert T. Internal and external factors affecting the development of neuropathic pain in rodents. Is it all about pain? Pain Pract 2003; 3(4):326–342.
40. Yezierski RP, Liu S, Ruenes GL, Kajander KJ, Brewer KL. Excitotoxic spinal cord injury: behavioral and morphological characteristics of a central pain model. Pain 1998; 75(1):141–155.
41. Jourdan D, Ardid D, Eschalier A. Analysis of ultrasonic vocalisation does not allow chronic pain to be evaluated in rats. Pain 2002; 95(1–2):165–173.
42. Fox A, Eastwood C, Gentry C, Manning D, Urban L. Critical evaluation of the streptozotocin model of painful diabetic neuropathy in the rat. Pain 1999; 81(3):307–316.
43. Apfel SC. Diabetic neuropathy models: are they relevant? Drug Discov Today 2006; 3:397–402.
44. Horowitz SH. Venipuncture-induced neuropathic pain: the clinical syndrome, with comparisons to experimental nerve injury models. Pain 2001; 94(3):225–229.
45. Allen TJ, Cooper ME, Lan HY. Use of genetic mouse models in the study of diabetic nephropathy. Curr Diab Rep 2004; 4(6):435–440.
46. Mogil JS, Wilson SG, Wan Y. Assessing nociception in murine subjects. In: Kruger L, editor. Methods in Pain Research. Boca Raton: CRC Press, 2001: 11–39.
47. Hounsom L, Tomlinson DR. Does neuropathy develop in animal models? Clin Neurosci 1997; 4:380–389.
48. Vincent AM, Russell JW, Low P, Feldman EL. Oxidative stress in the pathogenesis of

diabetic neuropathy. Endocr Rev 2004; 25(4):612–628.

49. Ruggiero C, Ferrucci L. The endeavor of high maintenance homeostasis: resting metabolic rate and the legacy of longevity. J Gerontol A Biol Sci Med Sci 2006; 61(5):466–471.
50. Greenberg JA, Wei H, Ward K, Boozer CN. Whole-body metabolic rate appears to determine the rate of DNA oxidative damage and glycation involved in aging. Mech Ageing Dev 2000; 115(1–2):107–117.
51. Pittenger GL, Ray M, Burcus NI, McNulty P, Basta B, Vinik AI. Intraepidermal nerve fibers are indicators of small-fiber neuropathy in both diabetic and nondiabetic patients. Diabetes Care 2004; 27(8):1974–1979.
52. Kennedy WR, Wendelschafer-Crabb G, Johnson T. Quantitation of epidermal nerves in diabetic neuropathy. Neurology 1996; 47(4):1042–1048.
53. Lin X, Taguchi A, Park S, Kushner JA, Li F, Li Y et al. Dysregulation of insulin receptor substrate 2 in beta cells and brain causes obesity and diabetes. J Clin Invest 2004; 114(7): 908–916.
54. Leonelli E, Bianchi R, Cavaletti G, Caruso D, Crippa D, Garcia-Segura LM et al. Progesterone and its derivatives are neuroprotective agents in experimental diabetic neuropathy: a multimodal analysis. Neuroscience 2007; 144(4):1293–1304.
55. Lauria G, Lombardi R, Borgna M, Penza P, Bianchi R, Savino C et al. Intraepidermal nerve fiber density in rat foot pad: neuropathologic–neurophysiologic correlation. J Peripher Nerv Syst 2005; 10(2):202–208.
56. Bianchi R, Buyukakilli B, Brines M, Savino C, Cavaletti G, Oggioni N et al. Erythropoietin both protects from and reverses experimental diabetic neuropathy. Proc Natl Acad Sci U S A 2004; 101(3):823–828.
57. Obrosova IG, Drel VR, Oltman CL, Mashtalir N, Tibrewala J, Groves JT et al. Role of nitrosative stress in early neuropathy and vascular dysfunction in streptozotocin-diabetic rats. Am J Physiol Endocrinol Metab 2007; 293(6):E1645–E1655.
58. Oltman CL, Davidson EP, Coppey LJ, Kleinschmidt TL, Lund DD, Adebara ET et al. Vascular and neural dysfunction in Zucker diabetic fatty rats: a difficult condition to reverse. Diabetes Obes Metab 2008; 10(1):64–74.
59. Ziegler D, Wiefels K, Dannehl K, Gries FA. Effects of one year of near-normoglycemia on peripheral nerve function in type 1 (insulin-dependent) diabetic patients. Klin Wochenschr 1988; 66(9):388–396.
60. Cardinal JW, Allan DJ, Cameron DP. Differential metabolite accumulation may be the cause of strain differences in sensitivity to streptozotocin-induced beta cell death in inbred mice. Endocrinology 1998; 139(6): 2885–2891.
61. Bergen HT, Mizuno T, Taylor J, Mobbs CV. Resistance to diet-induced obesity is associated with increased proopiomelanocortin mRNA and decreased neuropeptide Y mRNA in the hypothalamus. Brain Res 1999; 851(1–2):198–203.
62. Yoon YW, Lee DH, Lee BH, Chung K, Chung JM. Different strains and substrains of rats show different levels of neuropathic pain behaviors. Exp Brain Res 1999; 129(2):167–171.
63. Mogil JS. The genetic mediation of individual differences in sensitivity to pain and its inhibition. Proc Natl Acad Sci U S A 1999; 96(14):7744–7751.
64. Simone DA, Marchettini P, Caputi G, Ochoa JL. Identification of muscle afferents subserving sensation of deep pain in humans. J Neurophysiol 1994; 72(2):883–889.
65. Laursen RJ, Graven-Nielsen T, Jensen TS, Arendt-Nielsen L. The effect of differential and complete nerve block on experimental muscle pain in humans. Muscle Nerve 1999; 22(11):1564–1570.
66. Graven-Nielsen T, Mense S, Arendt-Nielsen L. Painful and non-painful pressure sensations from human skeletal muscle. Exp Brain Res 2004; 159(3):273–283.
67. Randall LO, Selitto JJ. A method for measurement of analgesic activity on inflamed tissue. Arch Int Pharmacodyn Ther 1957, 111(4):409–419.
68. Wuarin-Bierman L, Zahnd GR, Kaufmann F, Burcklen L, Adler J. Hyperalgesia in spontaneous and experimental animal models of diabetic neuropathy. Diabetologia 1987; 30:653–658.
69. Zhuang H-X, Wuarin L, Fei Z-J, Ishii DN. Insulin-like growth factor (IGF) gene expression is reduced in neural tissues and liver from rats with non-insulin-dependent diabetes mellitus, and IGF treatment ameliorates diabetic neuropathy. J Pharmacol Exp Ther 1997; 283(1):366–374.
70. Tishler M, Smorodin T, Vazina-Amit M, Ramot Y, Koffler M, Fishel B. Fibromyalgia in diabetes mellitus. Rheumatol Int 2003; 23(4):171–173.
71. Dobretsov M, Romanovsky D, Stimers JR. Early diabetic neuropathy: triggers and mechanisms. World J Gastroenterol 2007; 13:175–191.

72. Torebjörk HE, Lundberg LER, LaMotte RH. Central changes in processing of mechanoreceptive input in capsaicin – induced secondary hyperalgesia in humans. J Physiol 1992; 448:765–780.
73. Gracely RH, Lynch SA, Bennett GJ. Painful neuropathy: altered central processing maintained dynamically by peripheral input. Pain 1992; 51:175–194.
74. Cervero F, Laird JM. Mechanisms of touch-evoked pain (allodynia): a new model. Pain 1996; 68(1):13–23.
75. Romanovsky D, Hastings SL, Stimers JR, Dobretsov M. Relevance of hyperglycemia to early mechanical hyperalgesia in streptozotocin-induced diabetes. J Peripher Nerv Syst 2004; 9(2):62–69.
76. Chen SR, Pan HL. Hypersensitivity of spinothalamic tract neurons associated with diabetic neuropathic pain in rats. J Neurophysiol 2002; 87(6):2726–2733.
77. Ilnytska O, Lyzogubov VV, Stevens MJ, Drel VR, Mashtalir N, Pacher P et al. Poly(ADP-ribose) polymerase inhibition alleviates experimental diabetic sensory neuropathy. Diabetes 2006; 55(6):1686–1694.
78. Hoybergs YM, Meert TF. The effect of low-dose insulin on mechanical sensitivity and allodynia in type I diabetes neuropathy. Neurosci Lett 2007; 417(2):149–154.
79. Calcutt NA, Jorge MC, Yaksh TL, Chaplan SR. Tactile allodynia and formalin hyperalgesia in streptozotocin-diabetic rats: effects of insulin, aldose reductase inhibition and lidocaine. Pain 1996; 68:293–299.
80. Khan GM, Chen SR, Pan HL. Role of primary afferent nerves in allodynia caused by diabetic neuropathy in rats. Neuroscience 2002; 114:291–299.
81. Calcutt NA, Freshwater JD, Mizisin AP. Prevention of sensory disorders in diabetic Sprague-Dawley rats by aldose reductase inhibition or treatment with ciliary neurotrophic factor. Diabetologia 2004; 47(4): 718–724.
82. Maneuf YP, Blake R, Andrews NA, McKnight AT. Reduction by gabapentin of K+-evoked release of [3H]-glutamate from the caudal trigeminal nucleus of the streptozotocin-treated rat. Br J Pharmacol 2004; 141(4): 574–579.
83. Li F, Obrosova IG, Abatan O, Tian D, Larkin D, Stuenkel EL et al. Taurine replacement attenuates hyperalgesia and abnormal calcium signaling in sensory neurons of STZ-D rats. Am J Physiol Endocrinol Metab 2005; 288(1):E29–E36.
84. Tanabe M, Murakami T, Ono H. Zonisamide suppresses pain symptoms of formalin-induced inflammatory and streptozotocin-induced diabetic neuropathy. J Pharmacol Sci 2008; 107(2):213–220.
85. Rashid MH, Inoue M, Bakoshi S, Ueda H. Increased expression of vanilloid receptor 1 on myelinated primary afferent neurons contributes to the antihyperalgesic effect of capsaicin cream in diabetic neuropathic pain in mice. J Pharmacol Exp Ther 2003; 306(2): 709–717.
86. Dias JP, Ismael MA, Pilon M, de Champlain J, Ferrari B, Carayon P et al. The kinin B1 receptor antagonist SSR240612 reverses tactile and cold allodynia in an experimental rat model of insulin resistance. Br J Pharmacol 2007; 152(2):280–287.
87. Christianson JA, Ryals JM, Johnson MS, Dobrowsky RT, Wright DE. Neurotrophic modulation of myelinated cutaneous innervation and mechanical sensory loss in diabetic mice. Neuroscience 2007; 145(1):303–313.
88. Kontinen VK, Meert TF. Predictive validity of neuropathic pain models in pharmacological studies with a behavioral outcome in the rat: a systematic review. In: Dostrovsky JO, Carr DL, Koltzenburg M, editors. Proceedings of the Tenth World Congress on Pain. Seattle: IASP Press, 2004: 489–498.
89. Ahlgren SC, White DM, Levine JD. Increased responsiveness of sensory neurons in the saphenous nerve of the streptozotocin-diabetic rat. J Neurophysiol 1992; 68(6):2077–2085.
90. Bove G. Mechanical sensory threshold testing using nylon monofilaments: the pain field's "tin standard". Pain 2006; 124(1–2):13–17.
91. Kramer HH, Rolke R, Bickel A, Birklein F. Thermal thresholds predict painfulness of diabetic neuropathies. Diabetes Care 2004; 27(10):2386–2391.
92. Courteix C, Eschalier A, lavarenne J. Streptozocin-induced diabetic rats: behavioural evidence for a model of chronic pain. Pain 1993; 53(1):81–88.
93. Cotter MA, Jack AM, Cameron NE. Effects of the protein kinase Cbeta inhibitor LY333531 on neural and vascular function in rats with streptozotocin-induced diabetes. Clin Sci (Lond) 2002; 103:311–321.
94. Forman LJ, Estilow S, Lewis M, Vasilenko P. Streptozocin diabetes alters immunoreactive β-endorphin levels and pain perception after 8 wk in female rats. Diabetes 1986; 35: 1309–1313.
95. Gabra BH, Benrezzak O, Pheng LH, Duta D, Daull P, Sirois P et al. Inhibition of type 1

diabetic hyperalgesia in streptozotocin-induced Wistar versus spontaneous gene-prone BB/Worchester rats: efficacy of a selective bradykinin B1 receptor antagonist. J Neuropathol Exp Neurol 2005; 64(9):782–789.
96. Lee JH, McCarty R. Pain threshold in diabetic rats: effects of good versus poor diabetic control. Pain 1992; 50(2):231–236.
97. Jagodic MM, Pathirathna S, Nelson MT, Mancuso S, Joksovic PM, Rosenberg ER et al. Cell-specific alterations of T-type calcium current in painful diabetic neuropathy enhance excitability of sensory neurons. J Neurosci 2007; 27(12):3305–3316.
98. Malcangio M, Tomlinson DR. A pharmacologic analysis of mechanical hyperalgesia in streptozotocin/diabetic rats. Pain 1998; 76: 151–157.
99. Sugimoto K, Rashid IB, Shoji M, Suda T, Yasujima M. Early changes in insulin receptor signaling and pain sensation in streptozotocin-induced diabetic neuropathy in rats. J Pain 2008; 9(3):237–245.
100. Chu PC, Lin MT, Shian LR, Leu SY. Alterations in physiologic functions and in brain monoamine content in streptozocin-diabetic rats. Diabetes 1986; 35(4):481–485.
101. Calcutt NA, Malmberg AB, Yamamoto T, Yaksh TL. Tolrestat treatment prevents modification of the formalin test model of prolonged pain in hyperglycemic rats. Pain 1994; 58:413–420.
102. Rutledge LP, Ngong JM, Kuperberg JM, Samaan SS, Soliman KF, Kolta MG. Dopaminergic system modulation of nociceptive response in long-term diabetic rats. Pharmacol Biochem Behav 2002; 74(1):1–9.
103. Apfel SC, Arezzo JC, Brownlee M, Fedoroff II, Kessler JA. Nerve growth factor administration protects against experimental sensory neuropathy. Brain Res 1994; 634:7–12.
104. Kamiya H, Zhang W, Sima AA. Degeneration of the golgi and neuronal loss in dorsal root ganglia in diabetic biobreeding/Worcester rats. Diabetologia 2006; 49(11): 2763–2774.
105. Kamiya H, Murakawa Y, Zhang W, Sima AA. Unmyelinated fiber sensory neuropathy differs in type 1 and type 2 diabetes. Diabetes Metab Res Rev 2005; 21(5):448–458.
106. Roane DS, Porter JR. Nociception and opioid-induced analgesia in lean (Fa/-) and obese (fa/fa) Zucker rats. Physiol Behav 1986; 38(2):215–218.
107. Sullivan KA, Hayes JM, Wiggin TD, Backus C, Su OS, Lentz SI et al. Mouse models of diabetic neuropathy. Neurobiol Dis 2007; 28(3):276–285.
108. Pabbidi RM, Yu SQ, Peng S, Khardori R, Pauza ME, Premkumar LS. Influence of TRPV1 on diabetes-induced alterations in thermal pain sensitivity. Mol Pain 2008; 4:9.
109. Davar G, Waikar S, Eisenberg E, Hattori M, Thalhammer JG. Behavioral evidence of thermal hyperalgesia in non-obese diabetic mice with and without insulin-dependent diabetes. Neurosci Lett 1995; 190(3):171–174.
110. Ohsawa M, Kamei J. Role of intracellular calcium in thermal allodynia and hyperalgesia in diabetic mice. Brain Res 1999; 833(2): 278–281.
111. Osei K, Rhinesmith S, Gaillard T, Schuster D. Impaired insulin sensitivity, insulin secretion, and glucose effectiveness predict future development of impaired glucose tolerance and type 2 diabetes in pre-diabetic African Americans: implications for primary diabetes prevention. Diabetes Care 2004; 27(6): 1439–1446.
112. Stevens MJ, Zhang W, Li F, Sima AA. C-peptide corrects endoneurial blood flow but not oxidative stress in type 1 BB/Wor rats. Am J Physiol Endocrinol Metab 2004; 287(3):E497–E505.
113. Yeomans DC, Pirec V, Proudfit HK. Nociceptive responses to high and low rates of noxious cutaneous heating are mediated by different nociceptors in the rat: behavioral evidence. Pain 1996; 68(1):133–140.
114. Cheppudira BP. Characterization of hind paw licking and lifting to noxious radiant heat in the rat with and without chronic inflammation. J Neurosci Methods 2006; 155(1):122–125.
115. Ochoa JL, Yarnitsky D. Mechanical hyperalgesias in neuropathic pain patients: dynamic and static subtypes. Ann Neurol 1993; 33(5):465–472.
116. Kastenbauer T, Irsigler P, Sauseng S, Grimm A, Prager R. The prevalence of symptoms of sensorimotor and autonomic neuropathy in type 1 and type 2 diabetic subjects. J Diabetes Complications 2004; 18(1):27–31.
117. Dutta A, Naorem S, Singh TP, Wangjam K. Prevalence of peripheral neuropathy in newly diagnosed type 2 diabetics. Int J Diab Dev Countries 2005; 25:30–33.
118. Colin IM, Bouhassira D, Weiss K, Matthys K, Mathieu C, Van Acker K. Prevalence of polyneuropathy and neuropathic pain in type 1 and 2 diabetic patients and association with metabolic syndrome parameters. Diabetes 2007; 56(1 Suppl):A209.
119. Harris M, Eastman R, Cowie C. Symptoms of sensory neuropathy in adults with NIDDM in the U.S. population. Diabetes Care 1993; 16(11):1446–1452.

120. Maser RE, Steenkiste AR, Dorman JS, Nielsen VK, Bass EB, Manjoo Q et al. Epidemiological correlates of diabetic neuropathy. Report from Pittsburgh Epidemiology of Diabetes Complications Study. Diabetes 1989; 38(11):1456–1461.
121. Comi G, Canal N, Lozza L, Beccaria L, Meschi F, Vanini R et al. Peripheral nerve abnormalities in newly-diagnosed diabetic children. Acta Diabetol Lat 1986; 23(1):69–75.
122. Partanen J, Niskanen L, Lehtinen J, Mervaala E, Siitonen O, Uusitupa M. Natural history of peripheral neuropathy in patients with non-insulin-dependent diabetes mellitus. N Engl J Med 1995; 333(2):89–94.
123. Cheng WY, Jiang YD, Chuang LM, Huang CN, Heng LT, Wu HP et al. Quantitative sensory testing and risk factors of diabetic sensory neuropathy. J Neurol 1999; 246(5):394–398.
124. Boru UT, Alp R, Sargin H, Kocer A, Sargin M, Luleci A et al. Prevalence of peripheral neuropathy in type 2 diabetic patients attending a diabetes center in Turkey. Endocr J 2004; 51(6):563–567.
125. Tesfaye S, Chaturvedi N, Eaton SE, Ward JD, Manes C, Ionescu-Tirgoviste C et al. Vascular risk factors and diabetic neuropathy. N Engl J Med 2005; 352(4):341–350.
126. Sibal L, Law HN, Gebbie J, Home P. Cardiovascular risk factors predicting the development of distal symmetrical polyneuropathy in people with type 1 diabetes: a 9-year follow-up study. Ann N Y Acad Sci 2006; 1084:304–318.
127. American Diabetes Association. Standards of medical care in diabetes. Diabetes Care 2008; 31(1 Suppl):S12–S54.
128. Isomaa B, Almgren P, Henricsson M, Taskinen MR, Tuomi T, Groop L et al. Chronic complications in patients with slowly progressing autoimmune type 1 diabetes (LADA). Diabetes Care 1999; 22(8):1347–1353.
129. Baum P, Hermann W, Verlohren HJ, Wagner A, Lohmann T, Grahmann F. Diabetic neuropathy in patients with "latent autoimmune diabetes of the adults" (LADA) compared with patients with type 1 and type 2 diabetes. J Neurol 2003; 250(6):682–687.
130. Parving HH, Hommel E, Mathiesen E, Skott P, Edsberg B, Bahnsen M et al. Prevalence of microalbuminuria, arterial hypertension, retinopathy and neuropathy in patients with insulin dependent diabetes. Br Med J (Clin Res Ed) 1988; 296(6616):156–160.
131. Chao CC, Hsieh SC, Yang WS, Lin YH, Lin WM, Tai TY et al. Glycemic control is related to the severity of impaired thermal sensations in type 2 diabetes. Diabetes Metab Res Rev 2007; 23(8):612–620.
132. Claus D, Mustafa C, Vogel W, Herz M, Neundorfer B. Assessment of diabetic neuropathy: definition of norm and discrimination of abnormal nerve function. Muscle Nerve 1993; 16(7):757–768.
133. Sima AAF, Shafrir E editors. Animal Models of Diabetes. Amsterdam: Harwood Academic Publishers, 2001: 109–128.
134. Mordes JP, Desemone J, Rossini AA. The BB rat. Diabetes Metab Rev 1987; 3(3):725–750.
135. Pederson RA, Curtis SB, Chisholm CB, Gaba NR, Campos RV, Brown JC. Insulin secretion and islet endocrine cell content at onset and during the early stages of diabetes in the BB rat: effect of the level of glycemic control. Can J Physiol Pharmacol 1991; 69(8): 1230–1236.
136. Kanazawa M, Ikeda J, Sato J, Natoya Y, Ito H, Komeda K et al. Alteration of insulin and glucagon secretion from the perfused BB rat pancreas before and after the onset of diabetes. Diabetes Res Clin Pract 1988; 5(3): 201–204.
137. Reddy S, Bibby NJ, Fisher SL, Elliott RB. Longitudinal study of first phase insulin release in the BB rat. Diabetologia 1986; 29(11):802–807.
138. Svenningsen A, Dyrberg T, Markholst H, Binder C, Lernmark A. Insulin release and pancreatic insulin is reduced in young prediabetic BB rats. Acta Endocrinol (Copenh) 1986; 112(3):367–371.
139. Gabra BH, Sirois P. Hyperalgesia in non-obese diabetic (NOD) mice: a role for the inducible bradykinin B1 receptor. Eur J Pharmacol 2005; 514(1):61–67.
140. Hallfrisch J, Cohen L, Reiser S. Effects of feeding rats sucrose in a high fat diet. J Nutr 1981; 111(3):531–536.
141. Kim JY, Nolte LA, Hansen PA, Han DH, Kawanaka K, Holloszy JO. Insulin resistance of muscle glucose transport in male and female rats fed a high-sucrose diet. Am J Physiol 1999; 276(3 Pt 2):R665–R672.
142. Dobrian AD, Davies MJ, Prewitt RL, Lauterio TJ. Development of hypertension in a rat model of diet-induced obesity. Hypertension 2000; 35(4):1009–1015.
143. Srinivasan K, Viswanad B, Asrat L, Kaul CL, Ramarao P. Combination of high-fat diet-fed and low-dose streptozotocin-treated rat: a model for type 2 diabetes and pharmacological screening. Pharmacol Res 2005; 52(4):313–320.
144. Erdei N, Toth A, Pasztor ET, Papp Z, Edes I, Koller A et al. High-fat diet-induced reduction

in nitric oxide-dependent arteriolar dilation in rats: role of xanthine oxidase-derived superoxide anion. Am J Physiol Heart Circ Physiol 2006; 291(5):H2107–H2115.

145. Buettner R, Parhofer KG, Woenckhaus M, Wrede CE, Kunz-Schughart LA, Scholmerich J et al. Defining high-fat-diet rat models: metabolic and molecular effects of different fat types. J Mol Endocrinol 2006; 36(3):485–501.
146. Budohoski L, Challiss RA, Cooney GJ, McManus B, Newsholme EA. Reversal of dietary-induced insulin resistance in muscle of the rat by adenosine deaminase and an adenosine-receptor antagonist. Biochem J 1984; 224(1):327–330.
147. El Midaoui A, Ongali B, Petcu M, Rodi D, de Champlain J, Neugebauer W et al. Increases of spinal kinin receptor binding sites in two rat models of insulin resistance. Peptides 2005; 26(8):1323–1330.
148. Kava R, Greenwood MRC, Johnson PR. Zucker (fa/fa) rat. Ilar News 1990; 32:4–8.
149. Griffen SC, Wang J, German MS. A genetic defect in beta-cell gene expression segregates independently from the fa locus in the ZDF rat. Diabetes 2001; 50(1):63–68.
150. Corsetti JP, Sparks JD, Peterson RG, Smith RL, Sparks CE. Effect of dietary fat on the development of non-insulin dependent diabetes mellitus in obese Zucker diabetic fatty male and female rats. Atherosclerosis 2000; 148(2):231–241.
151. Durham HA, Truett GE. Development of insulin resistance and hyperphagia in Zucker fatty rats. Am J Physiol Regul Integr Comp Physiol 2006; 290(3):R652–R658.
152. Tirabassi RS, Flanagan JF, Wu T, Kislauskis EH, Birckbichler PJ, Guberski DL. The BBZDR/Wor rat model for investigating the complications of type 2 diabetes mellitus. ILAR J 2004; 45(3):292–302.
153. Poirier B, Lannaud-Bournoville M, Conti M, Bazin R, Michel O, Bariety J et al. Oxidative stress occurs in absence of hyperglycaemia and inflammation in the onset of kidney lesions in normotensive obese rats. Nephrol Dial Transplant 2000; 15(4):467–476.
154. Oltman CL, Coppey LJ, Gellett JS, Davidson EP, Lund DD, Yorek MA. Progression of vascular and neural dysfunction in sciatic nerves of Zucker diabetic fatty and Zucker rats. Am J Physiol Endocrinol Metab 2005; 289(1):E113–E122.
155. Weis J, Dimpfel W, Schroder JM. Nerve conduction changes and fine structural alterations of extra- and intrafusal muscle and nerve fibers in streptozotocin diabetic rats. Muscle Nerve 1995; 18(2):175–184.
156. Russell JW, Sullivan KA, Windebank AJ, Herrmann DN, Feldman EL. Neurons undergo apoptosis in animal and cell culture models of diabetes. Neurobiol Dis 1999; 6:347–363.
157. Sima AAF. Peripheral neuropathy in the spontaneously diabetic BB-Wistar-rat. An ultrastructural study. Acta Neuropathol (Berl) 1980; 51:223–227.
158. Sima AAF, Ristic H, Merry A, Kamijo M, Lattimer SA, Stevens MJ et al. Primary preventive and secondary interventionary effects of acetyl-l-carnitine on diabetic neuropathy in the bio-breeding Worcester rat. J Clin Invest 1996; 97(8):1900–1907.
159. Courteix C, lavarenne J, Eschalier A. RP-67580, a specific tachykinin NK_1 receptor antagonist, relieves chronic hyperalgesia in diabetic rats. Eur J Pharmacol 1993; 241:267–270.
160. Courteix C, Bardin M, Chantelauze C, lavarenne J, Eschalier A. Study of the sensitivity of the diabetes-induced pain model in rats to a range of analgesics. Pain 1994; 57(2):153–160.
161. Daulhac L, Mallet C, Courteix C, Etienne M, Duroux E, Privat AM et al. Diabetes-induced mechanical hyperalgesia involves spinal mitogen-activated protein kinase activation in neurons and microglia via *N*-methyl-d-aspartate-dependent mechanisms. Mol Pharmacol 2006; 70(4):1246–1254.
162. Ahlgren SC, Levine JD. Mechanical hyperalgesia in streptozotocin diabetic rats is not sympathetically maintained. Brain Res 1993; 616:171–175.
163. Courteix C, Bardin M, Massol J, Fialip J, lavarenne J, Eschalier A. Daily insulin treatment relieves long-term hyperalgesia in streptozocin diabetic rats. Neuroreport 1996; 7(12):1922–1924.
164. Zhuang HX, Snyder CK, Pu SF, Ishii DN. Insulin-like growth factors reverse or arrest diabetic neuropathy: effects on hyperalgesia and impaired nerve regeneration in rats. Exp Neurol 1996; 140(2):198–205.
165. Sasaki T, Yasuda H, Maeda K, Kikkawa R. Hyperalgesia and decreased neuronal nitric oxide synthase in diabetic rats. Neuroreport 1998; 9:243–247.
166. Kim H, Sasaki T, Maeda K, Koya D, Kashiwagi A, Yasuda H. Protein kinase Cbeta selective inhibitor LY333531 attenuates diabetic hyperalgesia through ameliorating cGMP level of dorsal root ganglion neurons. Diabetes 2003; 52:2102–2109.
167. Chen SR, Pan HL. Antinociceptive effect of morphine, but not mu opioid receptor number, is attenuated in the spinal cord of diabetic rats. Anesthesiology 2003; 99(6):1409–1414.

Chapter 10

Animal Models of HIV-Associated Painful Sensory Neuropathy

Sonia K. Bhangoo, Lauren Petty, and Fletcher A. White

Abstract

Painful distal sensory neuropathy is the most common neurological complication of HIV1 infection. There are several neuropathic pain syndromes associated with the disease; however, the most common is a sensory neuropathy called HIV sensory neuropathy (HIV-SN). HIV-SN can be subdivided into subacute or chronic distal sensory polyneuropathy (DSP) and subacute antiretroviral-induced toxic neuropathy (ATN). Both forms involve sensory loss and neuropathic pain. DSP occurs in up to 7–35% of HIV1-infected individuals and upwards of 34% of children infected with HIV1, while ATN develops following highly active antiretroviral therapy (HAART) treatment in up to 52% of patients. The mechanisms of HIV-SN remain unclear; however, the advent of several models of HIV1-associated peripheral neuropathy is helping unlock the mysteries surrounding HIV-SN. This chapter describes the known pathology of HIV1 and the resulting neuropathy syndromes, including descriptions of the models used to study this particular type of neuropathic pain.

1. HIV1-Associated Peripheral Neuropathy

HIV is known to cause extensive damage to the nervous system, both central and peripheral. Although the virus can cause several different forms of neuropathy, a progressive, painful polyneuropathy affecting the feet and hands is often the first clinically apparent sign of HIV infection. These clinical signs in either a subacute or chronic form are collectively termed, distal sensory polyneuropathy (DSP). DSP associated with HIV1 can be classified according to the onset of symptoms during HIV infection, etiology, and whether the neuropathology is primarily axonal or represents a demyelinating neuropathy in the peripheral nervous system. These neuropathological complications of the HIV1 virus are quite perplexing and may include nonneuronal cells of the immune system such as T cells and

Chao Ma and Jun-Ming Zhang (eds.), *Animal Models of Pain*, Neuromethods, vol. 49,
DOI 10.1007/978-1-60761-880-5_10,

macrophages. Alternatively, it is entirely possible that HIV1 indirectly influences the nervous system. Possible indirect means include neuronal and/or glial interaction with the viral coat proteins, proinflammatory mediators, and neurotoxins that may be released by infected macrophages and microglia (1). While infection with the virus itself comes with an incidence of neuropathy, patients may also be more likely to become symptomatic following the initiation of antiretroviral therapy (2).Taken together, these cellular interactions may be central to our understanding of the etiology of peripheral neuropathy syndrome associated with HIV1 infection.

The effect of neuroinflammatory mediators in neuropathic pain syndromes have recently been under intense investigation (3). While the mechanisms of HIV1-associated peripheral neuropathy are still not well understood, recent advances using in vivo rodent models have started to elucidate some of the elements that may be central to the pathology associated with the chronic pain syndrome. This section details what is known about the disease including clinical findings and the pathology. We then outline how current rodent models are being used to further understand the mechanisms underlying painful neuropathy associated with HIV1 infection.

2. Clinical Findings

The incidence of HIV-SN varies widely due to the disparity in reporting mechanisms and diagnostic criteria (~1/3 of HIV1-infected individuals) (2, 4). Prior to the advent of highly active antiretroviral therapy (HAART), the risk factors for DSP were low CD4+ T-cell counts and viral load of the patient and the sensory symptomology ranging from numbness to severe burning pain. The sensory neuropathy symptoms follow a classic stocking-glove distribution, which is bilateral in nature. The symptoms are first reported in the distal appendages and then progress to include the legs.

The symptoms of DSP associated with HIV-SN do not differ drastically from neuropathic pain produced by HAART and nucleoside reverse transcriptase inhibitor (NRTI) use. The main diagnostic tool used in distinguishing this type of "induced" neuropathy is the time course. Patients often develop a neuropathy syndrome soon after beginning NRTI treatment. This neuropathy has a highly temporal relationship in that symptoms ease following the reduction in dose of the drugs (2).

3. Pathogenesis

The pathology of HIV-SN is quite distinct. Like other neuropathy syndromes, HIV-SN is marked by distal axonal degeneration in primarily small, unmyelinated fibers. Nerve biopsies have shown epidermal denervation as well. Early onset can also exhibit inflammatory demyelinating neuropathy (2, 5). The cause of this demyelination and/or axon degeneration is currently under study.

4. Etiology of HIV Infection

HIV1 infects its target cells via the binding of its coat protein, gp120, along with a coreceptor hCD4, to either the CCR5 or CXCR4 chemokine receptors. Chemokine receptors are part of the G-protein coupled receptor superfamily and are found on a wide variety of cells, including neurons in both the peripheral and central nervous systems (6). Chemokines and their receptors are responsible for a host of neuronal responses including neural stem cell migration, neural transmission, and disease. With the chemokine coreceptors being expressed on both nonneuronal as well as neuronal cell populations, it is feasible that HIV1 can exert effects on neurons through these receptors (7–10).

It is well known that painful neuropathies result from alterations in neuronal function. There are several studies that now show gp120 to be a factor in HIV1-associated hypersensitivity. One study found that gp120 injected into the hindpaw of the rat led to a mechanical allodynia (7). Furthermore, gp120, when applied to DRG neurons results in excitation and the release of substance P. The same neurons that can be excited by gp120 also express the TRPV1 receptor (7). Neuronal excitation, Substance P, and TRPV1 are the factors involved in the chronic pain mechanism and could be involved in the pathogenesis of HIV1-associated neuropathic pain. Alternatively, gp120 may also bind to chemokine receptors located on Schwann cells, resulting in the release of TNF-alpha (TNFα) (11). TNFα has the ability to initiate an inflammatory cascade of events that can lead to pain syndromes. Furthermore, TNFα has the ability to sensitize different ion channels, which can then alter neuronal activity and lead to peripheral and central sensitization (12).

In addition to the virus itself leading to neuropathy, drugs used for HIV1 treatments have also been shown to lead to painful neuropathy syndromes. Many NRTIs used in HIV1 treatment are known to be neurotoxic and lead to pain syndromes (13). The mechanism for NRTI associated neuropathy is relatively unknown. One possible mechanism that has been suggested for the deleterious effects of NRTIs is mitochondrial dysfunction.

Increased numbers of abnormal mitochondria in neuronal axons and Schwann cells and depletion of mitochondrial DNA have been found in NRTI-associated neuropathic pain (14). Studies have also shown that mitochondrial abnormalities have been linked to neuropathic pain symptoms (9, 13, 15).

5. Treatments

Neuropathic pain is not only a costly syndrome to treat, but can also severely diminish the quality of life of the patient. The paradox of neuropathic pain is that symptoms can range from complete numbness to spontaneous paresthesias, hyperalgesia, and allodynia (16). Because the symptoms vary a great deal between patients, there is no gold standard of treatment. Treatments must be tailored toward the symptoms reported by the individual, and the outcome is often insufficient in completely alleviating the pain. Treatment of HIV-associated distal sensory polyneuropathy (HIV-DSP) typically follows the World Health Organization (WHO) analgesic ladder of stepwise pain control. Initially, patients receive nonopioid analgesics in the form of NSAIDs. If the pain remains uncontrolled, three categories of medications are used to treat the pain symptomology: anticonvulsants, antidepressants, and opioids. While opioids are the strongest pharmaceuticals available to combat pain, they have been proven to be controversial and fairly ineffective in treating neuropathic pain (17). Indeed, some patients do report moderate pain relief with opioids; however, the side effects, such as tolerance, constipation, nausea, and sedation, often limit its use in long-term therapy (18).

More recently, nonopioid drugs such as gabapentin and pregabalin have entered the market and have been slightly more successful in easing painful symptoms; however, their efficacy and side effects do not make them ideal drug candidates either. The use of anticonvulsants is based on the premise that damage to the peripheral nervous system produces changes in neuronal excitability. For example, carbamazepine, a sodium channel blocker, which is a common treatment for trigeminal neuralgia, has adverse side effects including excessive sedation and ataxia (19). Another anticonvulsant, gabapentin, is thought to bind to the $\alpha 2\delta$ subunit of the voltage-gated N-type calcium ion channel in the central nervous system, but the exact mechanism is still unknown. While the drug has been effective in reducing pain sensations in patients, some studies have shown that pain was only reduced by 33%, and there were still unpleasant side effects (18). Its newer pharmacological partner, pregabalin, has improved on some of the side effects, but again, the exact mechanism of these drugs and why calcium channel manipulation leads to fewer pain symptoms is not completely clear.

Antidepressants have a reasonable level of response in patients but again with side effects such as constipation, cognitive impairment, and sedation, therapeutic use is problematic (20, 21). Furthermore, amitriptyline, which is one of the commonly prescribed drugs for neuropathic pain, cannot be used in the elderly due to its anticholinergic side effects. Some of these nonopioid drugs have had moderate success; however, many times these drugs need to be used concomitantly with other drugs in the opioid category, and they fail to show an acceptable efficacy (22). With no effective treatment or surgery available to ease the pain in the majority of the patient population, it is imperative that we understand the mechanism and study possible therapeutic targets for neuropathic pain.

6. Mechanisms of HIV1-Associated DSP

There are key phenomena studied when one examines the literature on neuropathic pain mechanisms. During chronic pain, changes in neuronal activity, ion channel expression, immune cell activation, and gene transcription occur in the peripheral nervous system, spinal cord, and brain. Although the exact mechanism can vary depending on the specific type of pain, neuropathic pain following peripheral nerve injury is a consequence of enhanced excitability associated with the chronic sensitization of nociceptive neurons in the peripheral and central nervous system. This state of excitability could be due to a number of factors including, but not limited to, physical trauma, viral infection, inflammation, and cytotoxic drugs.

As noted previously, there are two ways in which HIV1 induced DSP may occur. The first manner is via viral shedding of HIV gp120. gp120 may indirectly produce DSP via glial/neuronal signaling in the DRG and/or spinal cord. The second manner in which DSP develops is through the direct activation of sensory neurons by gp120. Support for the indirect effect of gp120 includes the observation that perineural gp120 exposure is accompanied by nerve pathology (decreased fiber density and axonal swelling) and an upregulation of proinflammatory cytokines (8).

Complicating matters, HIV1 positive patients who are treated with HAART agents can also develop a painful sensory neuropathy. Intriguingly, the symptoms of this syndrome are clinically indistinguishable from those of HIV-induced distal symmetrical polyneuropathy, including a burning sensation in the hands and feet and hypersensitivity to pain. The fact that the two syndromes are usually seen in association with one another makes diagnosis more difficult.

To date, several methods have been used to model HIV1-associated DSP. With HIV1 neuropathy, it is important to consider what effects the virus itself has on the nervous system, in addition to the deleterious effects of neurotoxic NRTIs. Quite obviously, NRTIs will only be utilized to treat the individual once the HIV1 antibody test detected the presence of HIV1 in serum, saliva, or urine. However, on rare occasions, individuals with occupational exposure to HIV1 may receive postexposure prophylaxis in the form of NRTIs. Therefore, modeling of HIV-DSP may necessitate two insults to the recipient cells or tissue.

7. NRTIs

As discussed previously, patients are highly likely to develop a painful neuropathy syndrome after beginning the antiretroviral treatment. It is thus useful to mimic this treatment paradigm in rodents as well. Numerous studies have now shown that treating rodents with various antiretroviral drugs results in a severe neuropathic pain syndrome. Administration of NRTI drugs such as ddC, d4T, and ddI result in a dose-dependent mechanical and thermal hypersensitivity. For this model, rodents can be given a single intraperitoneal (i.p.) injection of the NRTI, with lasting results. It has also been shown that administration of the antiretroviral drugs intravenously (tail vein) results in a pain syndrome. For all methods of administration, doses ranging from 10 to 50 mg/kg are effective in the development of a neuropathic pain syndrome. Drugs are best when freshly prepared in saline on the day of the injection. Mechanical and thermal hyperalgesia are seen within 3 days of a single i.p. injection (9, 13, 23). Control animals given injections of saline show no change in mechanical or thermal hypersensitivity. Axonal pathology is also seen with the administration of these neurotoxic drugs. Increased myelin sheath thickness and Remak bundle degeneration are also observed as soon as 3 days postinjection. This model is relatively simple and consistent in leading to a neuropathy syndrome.

8. Viral Coat Protein gp120

In order to mimic what is seen clinically, a new model has been developed to utilize the gp120 protein. This model has been used in several papers now with consistent results (1, 10, 23). For the gp120 paradigm, the protein is delivered peripherally to the sciatic nerve, thus exposing the animals to the detrimental effects of the HIV1 virus particle. The animal is first anesthetized, after which an incision is made at the midthigh level to expose the sciatic nerve. For gp120 experiments, rgp120 HIV1 IIIB dissolved in buffered sterile saline is administered via a perineural surgical sponge.

An additional component of the HIV1 complex that can be added to this paradigm is the hCD4 glycoprotein coreceptor for gp120. The reason for the addition of this protein is to ensure that the gp120 binding mechanism occurs through the CXCR4 chemokine receptor. This addition is only necessary if the receptor is vital to the studies; however, this neuropathic pain model has been used without the addition of hCD4 (24). Although this combinatorial model adds a greater degree of complexity to the paradigm, it does provide for synergistic effects of the viral coat protein and HAART treatment to take place in vivo. These studies have shown the reduction in thermal and mechanical thresholds, and reduced intraepidermal nerve fiber density. Furthermore, with this model, an influx of macrophages is seen at the site of injury at the sciatic nerve. The importance of macrophages in this combinatorial paradigm is the possible source of proinflammatory mediators that could also contribute to the neuropathy symptoms.

9. Other Models

A model using transgenic technology has also been useful for a HIV1-associated neuropathic pain syndrome. A study by Keswani and colleagues (5) demonstrated that transgenic mice expressing the HIV coat protein gp120 under a GFAP promoter express the viral protein in both the CNS and PNS. The animals were then administered the NRT, ddI, orally through drinking water. The NRTI was dissolved in drinking water (5 mg/ml) for 4 weeks. Measurement of the daily dose was approximately 25 mg/day. This combination of treatments resulted in animals exhibiting many of the symptoms of HIV1-associated neuropathy. Animals display a distal, dying-back of small-fiber axons, which is similar to what is seen in HIV-SN patients. Other studies have looked at the effect of the gp120 coat protein on mechanical hypersensitivity in rodents as well. Several studies have looked at the effect of gp120 on pain hypersensitivity after intrathecal administration (25–29). In these studies, gp120 administration resulted in a rapid thermal and mechanical hypersensitivity, along with the release of proinflammatory cytokines such as TNFα and IL-1β. Other studies have looked at the role of glial activation and the release of proinflammatory mediators as part of the mechanism behind HIV1-associated peripheral neuropathy. Using transgenic animals and alternative methods of HIV delivery can answer questions about the peripheral neuropathy mechanism.

10. Summary

HIV-DSP is the most common neurological complication of HIV infection; approximately one-third of patients with HIV/AIDS experience symptoms due to HIV-DSP. Chronic pain associated with HIV can severely diminish the quality of life for a patient for significant periods of time. The models of HIV1-associated neuropathic pain outlined in this chapter provide a start to understanding not only how the virus itself may lead to peripheral neuropathy, but also the drugs used to treat the disease. With these novel models of HIV1-associated neuropathy, we start to see a picture how components of the HIV1 virus and the NRTIs may interact with respect to the development of chronic pain. As mentioned, gp120 and NRTIs alone have been shown to lead to a painful neuropathy. However, it is also important to take both components together when looking at the chronic pain mechanism. Studies doing this have shown that NRTI treatment in conjunction with the proalgesic gp120 protein can have a synergistic effect in the pain syndrome. These studies are preliminary in that they propose a method by which to study HIV1-associated neuropathic pain. While many of these studies have looked at the role of neuroinflammation and pain, these models can be used to dissect other aspects of the pain mechanism including the effect of gp120 and NRTIs on neuronal function, peripheral, and central sensitization. Using these models can further help our understanding of the disease and ways in which to improve the lives of those who suffer from it.

Acknowledgment

This work was supported by NIH grants at NINDS NS049136, NSO43095 and NIDA DA026040.

References

1. Herzberg U, Sagen J (2001) Peripheral nerve exposure to HIV viral envelope protein gp120 induces neuropathic pain and spinal gliosis. J Neuroimmunol 116:29–39.
2. Keswani SC, Pardo CA, Cherry CL, Hoke A, McArthur JC (2002) HIV-associated sensory neuropathies. AIDS 16:2105–2117.
3. Romero-Sandoval EA, Horvath RJ, DeLeo JA (2008) Neuroimmune interactions and pain: focus on glial-modulating targets. Curr Opin Investig Drugs 9:726–734.
4. Schifitto G, McDermott MP, McArthur JC, Marder K, Sacktor N, Epstein L, Kieburtz K (2002) Incidence of and risk factors for HIV-associated distal sensory polyneuropathy. Neurology 58:1764–1768.
5. Keswani SC, Jack C, Zhou C, Hoke A (2006) Establishment of a rodent model of HIV-associated sensory neuropathy. J Neurosci 26:10299–10304.
6. White FA, Feldman P, Miller RJ (2009) Chemokine signaling and the management of neuropathic pain. Mol Interv 9:188–195.

7. Oh SB, Tran PB, Gillard SE, Hurley RW, Hammond DL, Miller RJ (2001) Chemokines and glycoprotein120 produce pain hypersensitivity by directly exciting primary nociceptive neurons. J Neurosci 21:5027–5035.
8. Bhangoo S, Ren D, Miller RJ, Henry KJ, Lineswala J, Hamdouchi C, Li B, Monahan PE, Chan DM, Ripsch MS, White FA (2007) Delayed functional expression of neuronal chemokine receptors following focal nerve demyelination in the rat: a mechanism for the development of chronic sensitization of peripheral nociceptors. Mol Pain 3:38.
9. Bhangoo SK, Ren D, Miller RJ, Chan DM, Ripsch MS, Weiss C, McGinnis C, White FA (2007) CXCR4 chemokine receptor signaling mediates pain hypersensitivity in association with antiretroviral toxic neuropathy. Brain Behav Immun 21:581–591.
10. Bhangoo SK, Ripsch MS, Buchanan DJ, Miller RJ, White FA (2009) Increased chemokine signaling in a model of HIV1-associated peripheral neuropathy. Mol Pain 5:48.
11. Keswani SC, Polley M, Pardo CA, Griffin JW, McArthur JC, Hoke A (2003) Schwann cell chemokine receptors mediate HIV-1 gp120 toxicity to sensory neurons. Ann Neurol 54:287–296.
12. Miller RJ, Jung H, Bhangoo SK, White FA (2009) Cytokine and chemokine regulation of sensory neuron function. Handb Exp Pharmacol 194:417–449.
13. Joseph EK, Chen X, Khasar SG, Levine JD (2004) Novel mechanism of enhanced nociception in a model of AIDS therapy-induced painful peripheral neuropathy in the rat. Pain 107:147–158.
14. Dalakas MC (2001) Peripheral neuropathy and antiretroviral drugs. J Peripher Nerv Syst 6:14–20.
15. Flatters SJ, Bennett GJ (2006) Studies of peripheral sensory nerves in paclitaxel-induced painful peripheral neuropathy: evidence for mitochondrial dysfunction. Pain 122:245–257.
16. Devor M (2006) Sodium channels and mechanisms of neuropathic pain. J Pain 7:S3–S12.
17. Martin TJ, Eisenach JC (2001) Pharmacology of opioid and nonopioid analgesics in chronic pain states. J Pharmacol Exp Ther 299:811–817.
18. Rowbotham DJ, Peacock JE, Jones RM, Speedy HM, Sneyd JR, Morris RW, Nolan JP, Jolliffe D, Lang G (1998) Comparison of remifentanil in combination with isoflurane or propofol for short-stay surgical procedures. Br J Anaesth 80:752–755.
19. Eisenberg E, River Y, Shifrin A, Krivoy N (2007) Antiepileptic drugs in the treatment of neuropathic pain. Drugs 67:1265–1289.
20. McQuay HJ, Tramer M, Nye BA, Carroll D, Wiffen PJ, Moore RA (1996) A systematic review of antidepressants in neuropathic pain. Pain 68:217–227.
21. Kingery WS (1997) A critical review of controlled clinical trials for peripheral neuropathic pain and complex regional pain syndromes. Pain 73:123–139.
22. Lawson K (2008) Pharmacological treatments of fibromyalgia: do complex conditions need complex therapies? Drug Discov Today 13:333–340.
23. Wallace VC, Blackbeard J, Segerdahl AR, Hasnie F, Pheby T, McMahon SB, Rice AS (2007) Characterization of rodent models of HIV-gp120 and anti-retroviral-associated neuropathic pain. Brain 130:2688–2702.
24. Wallace VC, Blackbeard J, Pheby T, Segerdahl AR, Davies M, Hasnie F, Hall S, McMahon SB, Rice AS (2007) Pharmacological, behavioural and mechanistic analysis of HIV-1 gp120 induced painful neuropathy. Pain 133:47–63.
25. Milligan ED, Mehmert KK, Hinde JL, Harvey Jr. LO, Martin D, Tracey KJ, Maier SF, Watkins LR (2000) Thermal hyperalgesia and mechanical allodynia produced by intrathecal administration of the human immunodeficiency virus-1 (HIV 1) envelope glycoprotein, gp120. Brain Res 861:105–116.
26. Milligan ED, O'Connor KA, Nguyen KT, Armstrong CB, Twining C, Gaykema RP, Holguin A, Martin D, Maier SF, Watkins LR (2001) Intrathecal HIV-1 envelope glycoprotein gp120 induces enhanced pain states mediated by spinal cord proinflammatory cytokines. J Neurosci 21:2808–2819.
27. Spataro LE, Sloane EM, Milligan ED, Wieseler-Frank J, Schoeniger D, Jekich BM, Barrientos RM, Maier SF, Watkins LR (2004) Spinal gap junctions: potential involvement in pain facilitation. J Pain 5:392–405.
28. Ledeboer A, Gamanos M, Lai W, Martin D, Maier SF, Watkins LR, Quan N (2005) Involvement of spinal cord nuclear factor kappaB activation in rat models of proinflammatory cytokine-mediated pain facilitation. Eur J Neurosci 22:1977–1986.
29. Schoeniger-Skinner DK, Ledeboer A, Frank MG, Milligan ED, Poole S, Martin D, Maier SF, Watkins LR (2007) Interleukin-6 mediates low-threshold mechanical allodynia induced by intrathecal HIV-1 envelope glycoprotein gp120. Brain Behav Immun 21:660–667.

Chapter 11

Animal Models of Postoperative Pain

Chaoran Wu, Jun Xu, Sinyoung Kang, Christina M. Spofford, and Timothy J. Brennan

Abstract

Postoperative pain control remains difficult because the current treatments have limited efficacy; many patients experience moderate to severe pain after a variety of surgeries. Recognizing the gap between preclinical models of persistent pain and postsurgical pain, we and others have been interested in trying to better understand the mechanisms of pain caused by incisions, through the development of animal models. Plantar incision is one animal model for human postoperative pain. Models using hairy skin incision, gastrocnemius incision, and tail incision, and models for thoracotomy and abdominal surgery are reviewed. Relevant behaviors in relation to clinical postoperative pain are examined.

1. Introduction

When surveyed, most patients who undergo surgery report moderate to severe pain. Despite intensive management in the hospital, 20–30% of patients experience severe pain (1). Postoperative pain control is problematic because the current treatments have limited efficacy; many patients experience moderate to severe pain, especially pain with activities. Opioids are commonly used, yet they can cause many side effects including ileus, nausea, sedation, and respiratory depression. Local anesthetic nerve blocks are short-lived, and nonsteroidal anti-inflammatory drugs are helpful but can also be problematic.

Effective postoperative pain management decreases morbidity after surgery (2). In order to continue to reduce perioperative morbidity and health-care costs, it is critical that we continue to improve postoperative pain management. This will require that we develop a better understanding of the mechanisms of surgical pain. Those working in acute pain management appreciate that the etiology of incisional pain may be more different than that of inflammatory pain, chemical irritation, and nerve injury and that

Chao Ma and Jun-Ming Zhang (eds.), *Animal Models of Pain*, Neuromethods, vol. 49,
DOI 10.1007/978-1-60761-880-5_11, © Springer Science+Business Media, LLC 2011

the responses to treatments may also differ. Recognizing the disparity between preclinical models of persistent pain and postsurgical pain, we and others have been interested in trying to better understand the mechanisms of pain caused by incisions through the development of animal models.

In this chapter, we evaluate studies on incisional postoperative pain models in rats. By summarizing the incisional pain models, it is our hope that researchers will receive a focused and concise chapter on the models for postoperative pain. The chapter emphasizes the plantar incision model for postoperative pain. We also describe postoperative pain models using hairy-skin incision, hindlimb incision, laparotomy, thoracotomy, and tail incision.

2. Plantar Incision Model

Our laboratory began to develop and characterize the first rat model for human postoperative incisional pain (3). This model has several unique properties compared to other animal models of pain; importantly, it is caused by an incision, and profound and persistent pain-related behaviors are evident.

2.1. Acclimation and Surgery

We have performed our experiments on adult male rats. Rats undergo a variety of behavioral measures to assess nociceptive responses before and after incisions (3). It is strongly recommended that the experimenter acclimates the animals for 1–3 days in the testing room prior to starting the experiment, to minimize stress and exploratory activity. During this acclimation process, typically, the rats rest and develop minimal exploratory behavior to the clean testing environment. Subsequently, behavioral testing for nociception can be undertaken a few hours to a couple of days before any surgical procedure occurs; thus, rigorous baseline measurements are made after acclimation and before surgery. This reduces variability of nociceptive tests, likely by reducing the number of animals required to perform a study and perhaps by improving the validity of the data acquired. Our laboratory typically acclimates for 1–3 days before any procedure is undertaken, and attempts to obtain baseline activity either 1 or 2 days before surgical procedures are undertaken or the morning before animals undergo surgery.

Anesthesia is induced with a volatile anesthetic like isoflurane (1–2%) in either air or 100% oxygen in a sealed box. After induction, the same isoflurane concentration is delivered via a nose cone. The plantar aspect of either hindpaw is prepared in a sterile manner with a 10% povidone–iodine solution or any other sterile preparation, and the paw is placed through a hole in a sterile cloth or disposable plastic drape (Fig. 1a). A 1-cm longitudinal incision

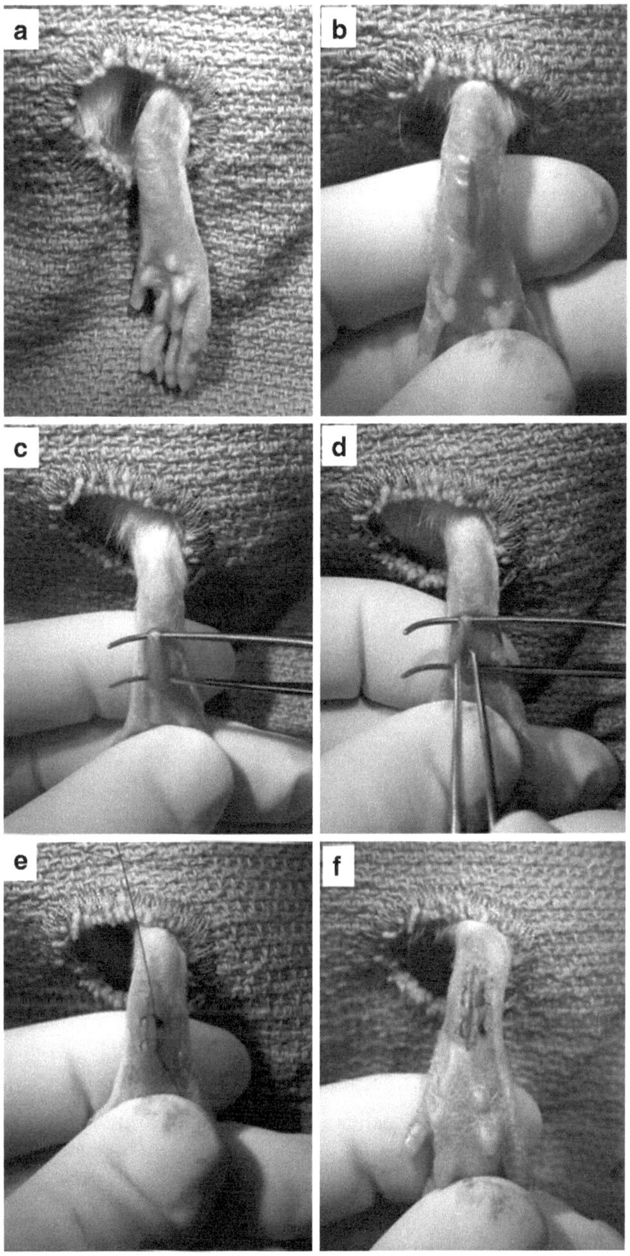

Fig. 1. Photographs of the rat plantar incision. (**a**) Preparation before incision. (**b**) A 1-cm longitudinal incision is made through the skin and fascia starting 0.5 cm from the proximal edge of the heel and extending toward the middle of the paw. (**c**) The underlying flexor muscle is elevated. (**d**) The muscle is split with a blunt dissection longitudinally. (**e**) After hemostasis, the wound is opposed with two mattress sutures of 5-0 nylon. (**f**) The completed incision.

is made with a number 11 blade, through skin and fascia of the plantar aspect of the paw (Fig. 1b), starting 0.5 cm from the proximal edge of the heel and extending toward the digits. In most animals, the flexor digitorum brevis muscle is elevated and incised

(Fig. 1c, d). The muscle origin and insertion remain intact. After hemostasis with gentle pressure, the skin is apposed with two mattress sutures of 5-0 nylon usually on an FS-2 needle (Fig. 1e, f). The wound site is covered with a mixture of polymixin B, neomycin, and bacitracin ointment. The sutures are removed under brief isoflurane anesthesia at the end of postoperative day 2. Typically, the wounds are well healed within 6–7 days.

After surgery, the rats are housed individually or in pairs on clean bedding consisting of organic cellulose fibers rather than wood chips. The incisions are checked daily for any sign of wound infection, dehiscence, or hematoma which would exclude rats from further study. Hematoma formation is uncommon and would typically be evident as a blue discoloration of the paw. We make every attempt to avoid infection. Although this is extremely rare, it could potentially confound the assessment of pain behaviors. Dehiscence can occur if the rats remove the suture during emergence or in the early postoperative period. Poor apposition of the wound could prolong the time course or intensify pain behaviors.

Recovery from anesthesia usually occurs in 15–20 min. Limping and some guarding occur during the recovery period. There is little evidence for severe pain and distress since spontaneous pain behaviors like vocalization or persistent flinching behavior are rarely observed. One to two hours are allowed for recovery before any behavioral observations are made, although in some cases pain-related behaviors have been examined as early as 15 min after the incision and completion of surgery. Importantly, use of a single drug like isoflurane minimizes anesthetic effects and permits testing early after surgery because pain behaviors are greatest in the early postoperative period.

2.2. Nociception to Mechanical Stimuli

Following emergence and recovery from anesthesia, unrestrained rats are placed beneath a clear plastic chamber (21 × 27 × 15 cm) on an elevated plastic (ADPI Enterprises, Philadelphia, PA) or stainless-steel mesh floor and allowed to reacclimate to the testing environment (3). Restraint should be minimized so that stress does not impact behavioral testing. Withdrawal responses to mechanical stimulation are determined using calibrated Semmes Weinstein monofilaments (Stoelting, Wood Dale, IL). These are nylon monofilaments with similar lengths and varying diameters, as the diameter of the filament increases when the force necessary to bend the filament is greater. For example, monofilaments with bending forces of approximately 14, 30, 42, 65, 73, 98, 149, 265, and 520 mN (milliNewtons) can be used. These monofilaments are applied individually from beneath the cage through the mesh floor (grid 12 × 12 mm) to an area adjacent to the wound (Fig. 2). Most often, we apply these just medial to the incision near the calcaneous. In our laboratory, each filament

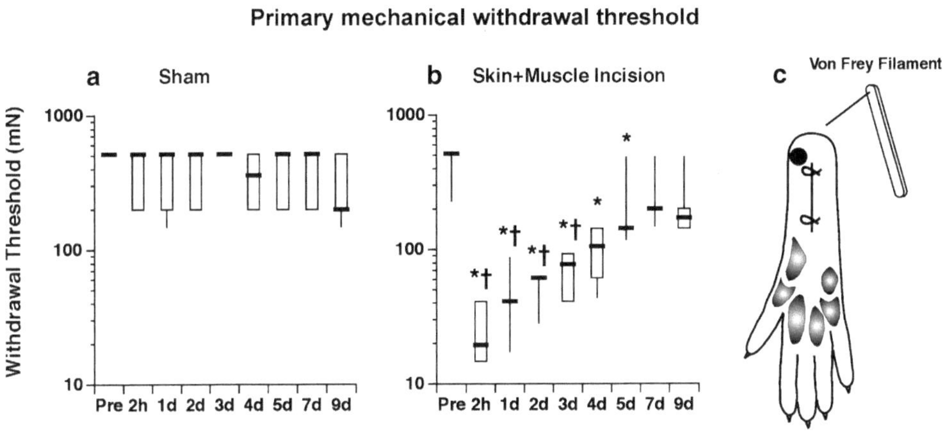

Fig. 2. Primary punctate withdrawal threshold after plantar incision. (**a**) Withdrawal thresholds in sham operated rats. (**b**) Withdrawal thresholds in rats after skin, fascia, and muscle incision. The forces, in milliNewtons (mN), are expressed as medians (*solid bold line*) with first and third quartiles (*box*), and 10th and 90th percentiles (*vertical lines*). (**c**) The drawing depicts the rat hindpaw and plantar incision. The *filled circle* adjacent to the wound near the medial heel is the testing area. *Asterisk* indicates $P<0.05$ versus preincision values. *Dagger* indicates $P<0.05$ versus sham incision. Adapted from Zahn et al. (8) with permission.

is applied once starting with approximately 10–15 mN strength fiber and continuing in an ascending order until a withdrawal response occurs or 522 mN (the cutoff value) is reached. When these filaments are applied to the unincised, normal rat paw near the calcaneous, withdrawal is usually not elicited until the cutoff filament or next weakest filament is applied. The withdrawal responses are brisk following the punctate filament application and are usually not accompanied by vocalization. The rats can be more active after the withdrawal and require a few minutes to return to a restful state before further testing is possible. Typically, after a 5–10 min test-free period, each filament is again applied once starting with 15 mN until a withdrawal response is elicited. This is repeated for a third time 5 min later. The lowest force from the three tests producing a response is considered the withdrawal threshold. The cutoff value, 522 mN, is recorded even if there was no withdrawal response to this force. Typically, the withdrawal threshold in the incised paw decreases after incision to less than 100 mN for 2–3 days (Fig. 2). A gradual return toward preincision withdrawal thresholds occurs over the next 5–6 days.

Because the forces exerted by the filaments are not continuous, we use nonparametric statistics to compare responses (4). Friedman's ANOVA followed by a post-hoc test like Dunn's test is used if appropriate. In some cases, normalizing the data to a percent of maximum possible effect or percent of baseline has been utilized. Analyses of other tests are parametric statistics.

Other methodologies have been employed to characterize the mechanical responsiveness after hindpaw incision. The details of the protocol are described by Chaplan et al. (5). Briefly, median 50% threshold can also be determined using the up–down method. A filament with midrange force is chosen, and stimuli are applied in a consecutive manner, either in an ascending or descending series. If no paw withdrawal occurs to the first filament, the next stronger one is applied; in the event of paw withdrawal, the next weaker filament is chosen. The optimal threshold calculation by this method requires six responses in the vicinity of the 50% threshold. Calculated thresholds are again not continuous data; thus, these results should be analyzed using nonparametric methods (4).

Mechanical hyperalgesia has been measured using the paw pressure technique. For this technique, rats are acclimated to restraint and repeatedly tested for paw withdrawal using the analgesymeter (7200, Ugo Basile, Italy) before incision. Gradual increased pressure is applied to the paw using a rounded cylindrical probe. This test has been used extensively for pain testing in other models and less extensively in incised rats. Baseline preincision testing yields a withdrawal threshold of approximately 150 g; cutoff is set at 250 g, and the end point is taken as paw withdrawal (6). The reduction in the withdrawal threshold for the paw pressure assay was maximal 1 day postincision (48 g), as compared to baseline and was similar to the unincised group by postoperative day 7–9 (Fig. 3).

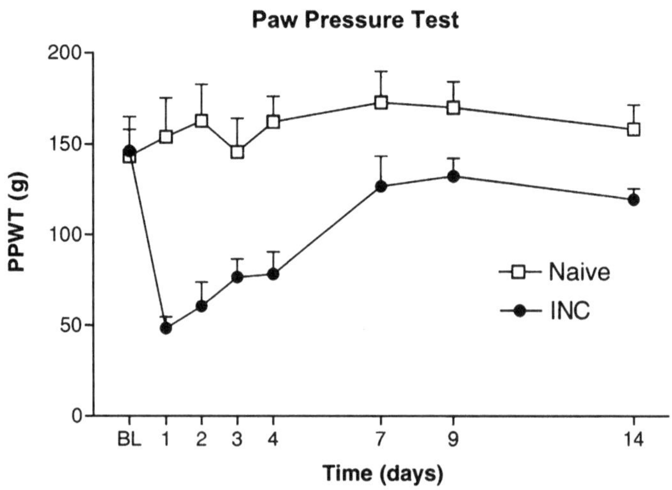

Fig. 3. Mechanical hyperalgesia, measured by Randall and Selitto analgesymeter, after incision of the plantar aspect of the rat hindpaw. Mechanical hyperalgesia was present throughout the time course tested. *BL* baseline, *INC* incision, *PPWT* paw pressure-withdrawal threshold. Values represent mean ± standard error of the mean. From Whiteside et al. (6) with permission.

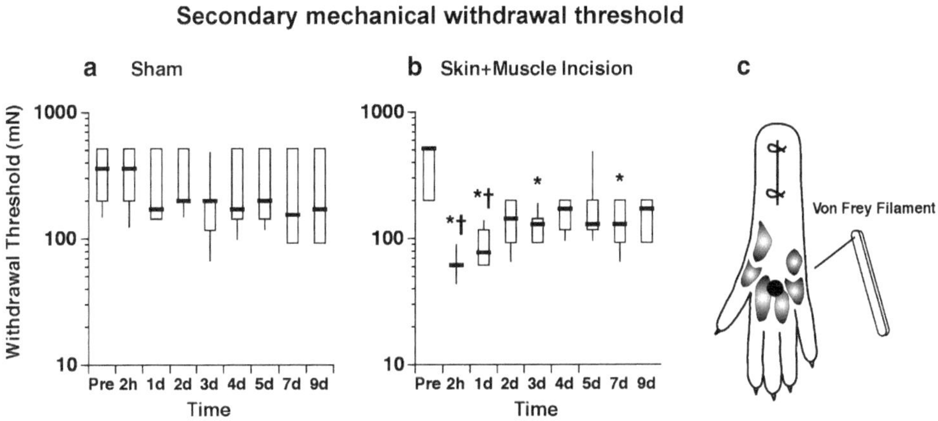

Fig. 4. Secondary (remote) punctate withdrawal threshold after plantar incision. (**a**) Remote withdrawal thresholds in sham operated rats. (**b**) Remote withdrawal thresholds in rats after skin, fascia, and muscle incision. The forces, in milliNewtons (mN), are expressed as medians (*solid bold line*) with first and third quartiles (*box*), and 10th and 90th percentiles (*vertical lines*). (**c**) The drawing depicts the rat hindpaw and plantar incision. The *filled circle* adjacent to the wound near the digits is the testing area. *Asterisk* indicates $P<0.05$ versus preincision values. *Dagger* indicates $P<0.05$ versus sham incision. Adapted from Zahn et al. (8) with permission.

2.2.1. Secondary Hyperalgesia

A reduced withdrawal threshold to punctate mechanical stimuli also occurs outside the incision in uninjured areas surrounding the incision (Fig. 4). Responses in uninjured territories are considered secondary hyperalgesia, a process that also occurs in patients after surgery (7). Pain in the secondary zone is caused by central nervous system sensitization because the sensory fibers function normally outside the area of injury. This test is particularly useful for studying aspects of central sensitization after incisions. Baseline median withdrawal thresholds to filaments applied between the distal tori, an area approximately 1 cm from the intended plantar incision, tend to be less than proximal test sites. Withdrawal thresholds from monofilaments applied 1 cm from the plantar incision were decreased 2 h and 1 day after surgery, indicating that the secondary hyperalgesia and central sensitization occur after plantar incision (8).

2.3. Nociception to Heat

The clinical significance of heat hyperalgesia is not well understood despite the popularity of its testing in basic science pain models. Heat hyperalgesia occurs in humans after surgery and is robust (9). Withdrawal latencies to heat are assessed by applying a focused radiant heat source in unrestrained rats. The heat stimulus is a focused light from a lamp, with a small aperture diameter of 6–8 mm, applied from beneath a heat-tempered glass floor (3-mm thick) on the middle of the incision (8). Some investigators use glass that is heated to skin temperature. Paw withdrawal latencies are measured to the nearest 0.1 s. Three trials, 5–10 min apart, are used to obtain an average paw withdrawal latency.

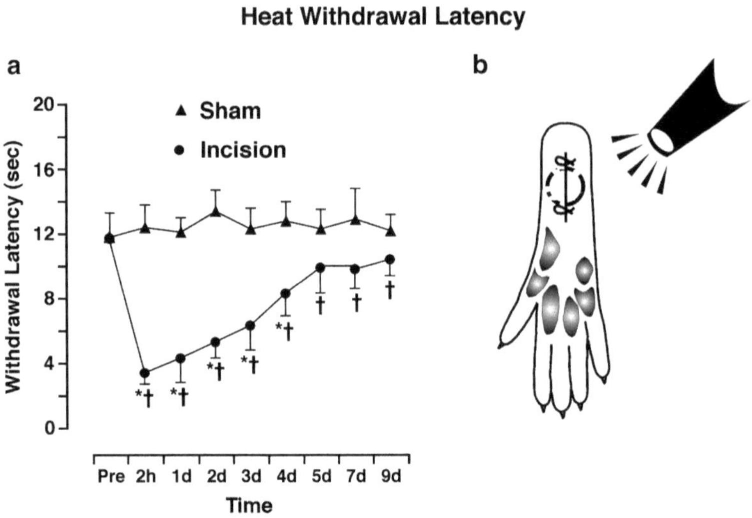

Fig. 5. Heat responses after plantar incision. (**a**) Heat withdrawal latencies in sham operated rats and in rats after skin, fascia, and muscle incision. Values represent mean ± standard error of the mean. (**b**) The drawing depicts the rat hindpaw and plantar incision. The *circle* overlying the incision is the testing area. *Asterisk* indicates $P < 0.05$ versus preincision values. *Dagger* indicates $P < 0.05$ versus sham. Adapted from Zahn et al. (8) with permission.

After plantar incision, the withdrawal latency to radiant heat applied directly to the wound decreased from 10 to 12 s to as low as 3 s (Fig. 5a–5b) a gradual return toward preincision values occurs. Withdrawal latency to radiant heat was tested at two areas remote from the incision, and no secondary heat hyperalgesia was observed (not shown). In some cases, we use a more gradual heat stimulus, one that elicits withdrawal with a greater latency (25 s) (10). In this case, the withdrawal latency after incision is also usually greater, approximately 5–10 s after incision. Thus far, evidence indicates both protocols effectively measure heat hyperalgesia.

2.4. Unprovoked Pain

2.4.1. Guarding

After plantar incision, there is a small amount of spontaneous pain behavior, such as biting, scratching, licking, or vocalization. We have observed that rats do not normally bear weight on the incision in the postoperative period. We therefore developed a cumulative pain score that utilizes the position in which both paws are positioned before and after incision (3, 11).

As with other pain behavior tests, unrestrained rats are placed beneath a clear plastic chamber on a smaller elevated plastic-mesh floor (grid 8 × 8 mm, ADPI Enterprises, Philadelphia, PA). Using an angled magnifying mirror, the incised and nonincised paws are viewed. Both paws are observed for a 1-min period every 5 min for 1 h. Depending on the position in which the paw is observed during the majority of the 1-min scoring period, a 0, 1, or 2 is given. Full weight bearing of the paw (wound is blanched or distorted by the mesh) is assigned a score = 0. If the paw is completely off the mesh, a score of 2 is recorded. If the area of the incision touches the mesh without blanching or distorting, a 1 is given. The 12 scores (0–24) obtained during the 1-h session for each paw are added, and the difference between scores from the incised paw and nonincised paws are the cumulative pain score for that 1-h period. This pain behavior is increased after incision, and a gradual return towards preincision values occurs over the next few days (Fig. 6).

2.4.2. Weight Bearing

Hindlimb weight bearing after plantar incision has been measured using an incapacitance meter (Stoelting, CA, USA), a pair of adjacent scales on which the rat stands. The incapacitance meter averages the weight over 3–5 s. Usually, three measurements are taken, and the mean percentage of weight on the incised paw is expressed as a percentage of the total weight on both hindpaws. Normal and sham operated rats distribute their weight equally on both hindpaws. The difference of weight distribution between an incised paw and nonincised paw has been suggested to reflect the nociception after incision.

After acclimation and baseline measurements, the rats undergo hindpaw incision (6). After 1, 3, 7, and 14 days postsurgery, measures for weight bearing are made. Incision of the plantar surface of the hindpaw produced a significant reduction in weight bearing on the incised limb 1 and 3 days later. The reduction in weight bearing was maximal 1 day postincision (Fig. 7).

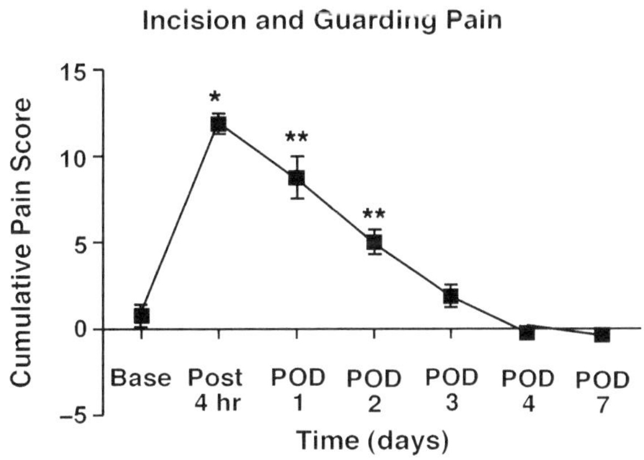

Fig. 6. Effect of incision on guarding pain behavior. Cumulative pain score after skin, fascia, and muscle incisions. $^{*}P<0.05$, $^{**}P<0.01$ versus baseline. Adapted from Wu et al. (28) with permission.

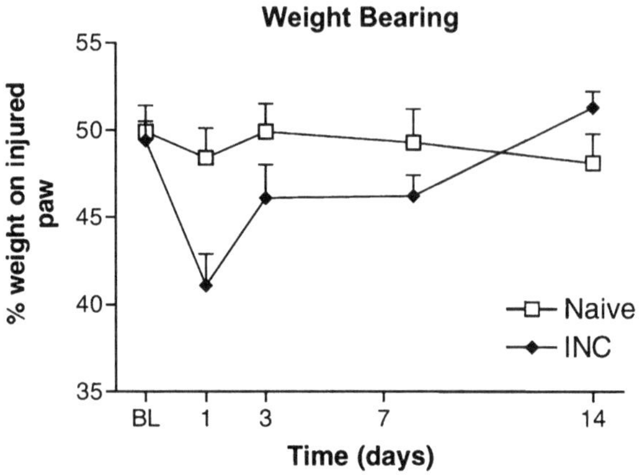

Fig. 7. Weight bearing, measured by an incapacitance meter, after incision of the plantar aspect of the rat hindpaw. A decrease in weight bearing of the incised limb was present 1 day after incision. *BL* baseline, *INC* incision group. Values represent mean ± standard error of the mean. From Whiteside et al. (6) with permission.

3. Hairy-Skin Incision

A model of hairy-skin incision has been described (12). The model contributes to our understanding of incisional pain mechanisms because previous models used glabrous-skin incisions, which are functionally different from hairy-skin incisions.

3.1. Preparation

Experiments are conducted in rats that have been handled daily (over 7–10 days) and are thus familiarized with the behavioral tester, the experimental environment, and the specific experimental procedures. The dorsal cutaneous muscle contractile response is abolished in the area to undergo incision and testing by repeated handling. The hair over the skin to be tested is shaved with electric shears 24 h before testing; local irritation produced by this clipping resolves overnight.

3.2. Testing

Under general anesthesia, a 1-cm longitudinal (parallel to the midline) incision of the skin is made in the back (Fig. 8). The wound is closed with one suture of 3-0 or 4-0 silk, and the rats are allowed to recover from anesthesia. Mechanical stimuli (monofilaments, 18–250 mN) are applied to measure nociception, indicated by twitching of local subcutaneous muscles, namely, the cutaneous trunci muscle reflex. Each von Frey filament is applied four times in any one trial, each application is spaced 2–3 s apart, and sequential von Frey filaments are applied in ascending order of stiffness. A graded response score, averaging the twitches weighted by their intensity of response or vigor, i.e., speed and rostrocaudal extent, is assessed for 3 days before and over 7 days

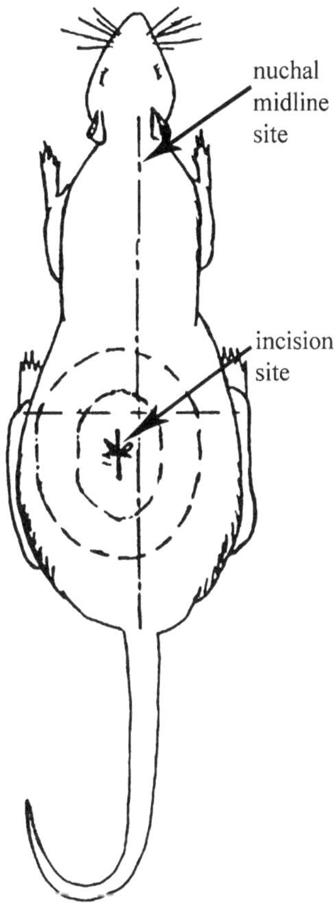

Fig. 8. Dorsum of the rat showing the location of the 1-cm-long skin incision, posterior to the L4 transverse process and 0.5 cm to the *left* of midline. The *broken vertical line* represents the midline and *broken circles* depict regions at 1 and 2 cm distance from the incision line. From Duarte et al. (12) with permission.

after incision. In the graded response analysis, a contraction similar to that from pinprick (5–7 cm of the skin contracts, along the rostrocaudal axis) is scored as 1.0, a weaker or shorter contraction (2–4 cm or less) is scored as 0.5, and no visible contraction is scored as 0 by the experimenter. Since four trials are applied for each von Frey filament, the maximum response that can be scored is 4.0. This can be decreased by units of 0.5 until there is no response. The raw scores are normalized by the maximum possible response (4.0), and the resulting scores given as a weighted average, the graded response, ranging from 0 to 1, calculated as follows: Graded response $= \Sigma v_i/4$ where v_i is the vigor of the cutaneous trunci muscle reflex to the *i*th stimulation and the sum is taken over all four stimuli for each von Frey filament (12). This visual assessment of the strength and extent of muscle contraction is subjective and semiquantitative, and the score assigned to the extent of contraction is dependent on learned judgment. To minimize potential contamination of data by subjective bias, the authors had only one investigator handle and subsequently test any one cohort of rats (usually numbering eight) for all the different procedures done in that cohort. Measured 0.5 cm from the wound, the graded responses are increased after incision as early as 30 min and peak responses are seen at 3–6 h, which remain increased for at least one full week, indicating primary mechanical hyperalgesia. These changes are transiently reversed by parenterally administered morphine. By comparison, secondary graded responses to the incision, measured at 2.0 cm from the wound, are less severe and briefer than primary changes.

The postoperative model by Duarte et al. (12) incorporated several new pain tests. First, the intensity of the response was measured, rather than a simple threshold. Second, the responses were separated into allodynic-like and hyperalgesic-like responses based upon careful preincision characterization of the responses. Finally, in this model, the regions of primary and secondary hypersensitivity are easily distinguished and these have distinctive time courses and unique responses to treatments. This permits a precise examination of these sensitization phenomena, primary (injured territories) and secondary (uninjured territories) responses, as well as hyperalgesia and allodynia.

4. Gastrocnemius Incision Model

As noted previously, it has been demonstrated that decreased threshold to mechanical stimulation is present in sites remote from surgical injury (for example, Fig. 4). Such secondary hyperalgesia is different than primary hyperalgesia and is considered to be a result of central sensitization (4). A rat gastrocnemius incision model was developed that uses hindpaw testing to study secondary hyperalgesia (13, 14).

Acclimation and preparation for surgery are similar to these procedures described for plantar incision. Under general anesthesia with isoflurane, a 3-cm longitudinal incision is made with a number 11 scalpel blade through the skin and fascia of the rat posterior hindleg beginning 1–1.5 cm from the edge of the heel and toward the popliteal region. After being separated from the cutaneous tissue, the underlying gastrocnemius muscle is bluntly split and retracted while the muscle insertion remains intact. Then, the incised skin is apposed with 3–4 nylon sutures.

Secondary hyperalgesia is examined after surgery by applying punctate mechanical stimuli to the plantar aspect of the hindpaw, which is distal from the gastrocnemius incision site. Only responsiveness to punctate mechanical stimulus is enhanced. No secondary heat hyperalgesia is present. Reduced withdrawal thresholds to punctate stimulus are observed 2 h after surgery, and these are sustained for 6–8 days. A separate group that only includes the incision in the skin and fascia without the surgical procedure on the muscle results in a transient increase in responsiveness to punctate mechanical stimulus. Pharmacologic studies support that the reduced withdrawal thresholds are pain-related responses.

5. Other Hindlimb Incisions

The findings from studies on a rat model of hindlimb stab wound injury are consistent with the data of the gastrocnemius incision model (15). The stab wound injury model includes two incisions, which are made 1 cm distal to the knee joint with a number 11 scalpel blade, one on the anterior and the other on the posterior aspect of the hindlimb. The incisions are 0.5-cm long and 1-cm deep through skin and underlying muscles. Secondary mechanical hyperalgesia is examined by measuring withdrawal response frequencies to punctate mechanical stimuli applied to the plantar hindpaw. After the stab wound, secondary heat hyperalgesia did not develop, while secondary mechanical hyperalgesia was present for 2–3 days. Such secondary mechanical hyperalgesia was not induced if incisions were made in skin only.

Previous data generated by investigators examining neuropathic pain models are in agreement with results from the gastrocnemius incisions. Decreased hindpaw withdrawal thresholds to punctate mechanical stimuli were an unexpected finding in a group of rats that received a sham surgery for unilateral sciatic nerve constriction (16). Reduced withdrawal threshold was present for 40 days after the surgical procedure that included an incision in the upper hindlimb, blunt dissection of the biceps femoris muscle, and exposure of the sciatic nerve. It has been suggested that this may represent a rat model for persistent postoperative pain. This more persistent pain may occur because there is manipulation of the sciatic nerve (16).

6. Laparotomy Models for Postoperative Pain

The initial models developed to study mechanisms specific to incisional and postoperative pain were laparotomy models that included an intraperitoneal incision and ovariohysterectomy. Laparotomy models have obvious clinical relevance to patient's abdominal and pelvic surgeries. Two abdominal surgery models are described in this chapter.

6.1. Ovariohysterectomy

In rats, a midline ovariohysterectomy is performed under general anesthesia (17, 18). The ovarian ligaments and cervix are ligated with 4-0 silk using the clamp technique. A continuous subcuticular/fascial closure is utilized, and four simple interrupted sutures are placed in the skin. The rats are allowed 35 min to recover fully before the first postoperative nociceptive tests are made. Mechanical nociceptive thresholds (paw pressure test) and thermal nociceptive thresholds (tail-flick latency) are then measured at stepwise increasing intervals for 480 min to study the development of central nervous system hypersensitivity. The paw pressure test was enhanced, indicating remote (secondary) mechanical hyperalgesia after ovariohysterectomy.

When measured 30–90 min after anesthesia, there was a significant increase in paw pressure thresholds above baseline in the animals subjected to anesthesia and surgery, without analgesia. The explanation for this change did not seem to be the remaining effects of anesthesia since this hypoalgesia was not observed in the rats subjected to anesthesia alone. The paw pressure thresholds then returned to baseline levels about 110 min after anesthesia was discontinued. From this point, thresholds decreased below baseline values and did not return to baseline before termination of the experiment.

Results of the tail-flick latency testing did not indicate the development of any significant remote heat hyperalgesia. The importance of remote hindlimb hyperalgesia after hysterectomy, however, may be less clinically relevant than direct incisional sensitivity. Remote hindlimb hyperalgesia does not occur in patients; when the pressure pain threshold was assessed in patients undergoing hysterectomy, increased tenderness to mechanical stimulation remote from the surgical wound at the anterior surface of the thigh and at the tibial tuberosity could not be demonstrated postoperatively (19).

6.2. Subcostal Incision

In another abdominal surgery model, exploratory locomotor behavior and conditioned operant responses are assessed in rats after a subcostal laparotomy and manipulation of the underlying viscera (20). Under pentobarbital anesthesia (50 mg/kg, i.p.), a diagonal 3-cm incision is made 0.5 cm below and parallel to the lowest rib on the left side, penetrating into the peritoneal cavity. The musculature is manipulated by inserting the index finger into the peritoneal cavity and stretching the abdominal wall. Approximately 10 cm of the small intestine is exteriorized and vigorously manipulated between the thumb and forefinger. The intestine is then placed inside the peritoneal cavity, and the wound is sutured in three layers consisting of the peritoneal lining, abdominal muscles, and skin. Ambulatory activity, stereotypy (small confined movements), and rearing are assessed at various times after surgery, using commercially available infrared monitoring system.

Laparotomy decreased ambulation and rearing by approximately 50%, 24 h after surgery, and stereotypic responses was affected to a lesser degree (20). Morphine (3 mg/kg) reversed the effects of surgery on ambulation and stereotypy, but not rearing, and the dose-effect curve for morphine (0.03–1 mg/kg) was shifted to the left by 5 mg/kg ketorolac. Subcostal laparotomy also decreased operant responses for food in rats undergoing subcostal incision. Infiltration of the wound with bupivacaine or denervation of the abdominal wall in the area of the incision produced a reversal of the effects of surgery on food-maintained behavior (21). In contrast, when the visceral organs are manipulated, alleviation of incisional pain by these methods produces little or no benefit behaviorally. These models have obvious relevance to a common clinical procedure, and the particular injury is different from a skin and muscle incision. Additional peritoneal and visceral injuries are likely associated with laparotomy and may intensify pain behaviors.

7. Thoracotomy Pain Models

In patients, thoracic surgery induces severe postoperative pain. Pain with coughing is severe, and pulmonary function is impaired. Chronic postthoracotomy pain, pain that persists at least 2 months after surgery, is also a significant clinical problem. The importance of the incision, intercostal nerve injury, and rib retraction on pain after thoracic surgery are not understood (22, 23). This summary will focus on the acute aspects of postthoracotomy pain.

Experimental animal models of postthoracotomy pain have been developed recently. Nara et al. developed the first model in rats (24). Male Wistar rats, 200–250 g, undergo isoflurane induction and maintenance, followed by shaving and skin asepsis. A 3-cm skin, fascia, muscle, and pleura incision is made in the posteriolateral thoracic spine. The intercostal nerves from T4 and T5 are isolated along with the posterior and lateral cutaneous nerve branches. The intercostal nerves are ligated loosely with two 4-0 chromic gut sutures 5 mm apart. The ribs are not traumatized. The incision is then closed in layers, and the pneumothorax is evacuated using a 20G needle catheter. Sham animals undergo a similar operative procedure, except the nerves are not ligated or damaged. In a third group of animals, a similar operation is performed, and the fourth and fifth ribs are cut with scissors, 1 cm distal to the head of the rib. In the group with rib cutting, the rib fragment is not removed, and the intercostal nerves are not damaged.

Pain behaviors are tested on days 1, 6, 13, 20, and 27 after surgery. Three tests are performed on each rat along the T4 and T5 dermatomes bilaterally: von Frey testing, forceps pinch, and cold sensitivity to acetone. All animals are acclimated for 15 min before pain behavior testing is done. The von Frey filaments are applied perpendicularly to the incision site and to the contralateral side. The recorded threshold is the filament that causes scratching behavior followed by licking. Pinch is tested by apposing the pointed end of forceps for 1 s along the back at the T4 and T5 dermatomal levels bilaterally. This pinch corresponds to 2 kg of force. In the cold stimulation test, 500 μl of 80% acetone is dripped from a syringe with a 20G Teflon needle 2 cm above the incision site or contralateral zone at the level of T4. The acetone is applied three times with 2 min between applications, and the number of scratching behaviors in 1 min is recorded.

Animals undergoing this operative procedure display cutaneous hypersensitivity to mechanical stimulus using von Frey filaments. The most robust responses were observed in the group that underwent incision plus nerve ligation. The three controls (nerve ligation with contralateral testing, sham operation, and rib cutting without nerve injury) also showed pain behavior but to a lesser degree. There were no significant differences or changes between the sham-incised side and contralateral side in the two control group operations at any times tested.

Animals undergoing nerve ligation also showed increased cold response to acetone through postoperative day 20. Unlike von Frey testing, there was little change when acetone was applied to the contralateral side in nerve ligation animals. Pinch scores were also significantly higher in the ligation group compared to sham group for the entire time course studied, up to 27 days.

This procedure has many similarities to patients with long-term postthoracotomy pain. Similar to humans, these animals display mechanical, pressure, and cold sensitivity. The time course for behavioral changes parallels other nerve injury models.

Although the model by Nara displays some of the neuropathic components seen in patients who have postthoracotomy pain, the intercostal nerves are rarely, if ever, ligated, raising concern about the initiating event in this model. To overcome some of these concerns, Buvanendran (25) revised the model from Nara and colleagues (24) of postthoracotomy pain in rats. This model utilized a similar surgery as the Nara model, but instead of nerve ligation used prolonged rib retraction, a common practice in humans undergoing thoracic surgery. Animals were tested for mechanical allodynia with von Frey filaments and cold sensitivity using acetone for 40 days postoperatively. In the animals that developed allodynia, those with the longest duration of rib retraction (60 min) displayed the most robust pain behaviors.

Only half of the rats undergoing this procedure will show signs of increased sensitivity with von Frey testing (25). Furthermore, in rats undergoing skin, muscle, and pleura incisions without rib retraction, only 10% will develop mechanical sensitivity to von Frey testing. The animals are all genetically and environmentally similar, which raises the question as to whether the surgical procedure and pathology is similar to humans undergoing throracic surgery. Perhaps other factors result in protection or susceptibility to post-thoracotomy pain.

Histological analysis of intercostal nerves after rib retraction revealed extensive Wallerian degeneration, similar to nerves in other neuropathic models. However, these histological changes were only apparent in animals displaying the most severe mechanical sensitivity. Rib retraction can also cause direct axonal compression as well as stretching. The authors found that the T5 nerve was normal in this model, suggesting that compression may be more important than stretch.

Animal models of thoracic surgery may be important for elucidation of pain mechanisms in patients that undergo this extensive and painful surgical procedure. Perhaps animal models of chronic postthoracotomy pain will facilitate the understanding of persistent postthoracotomy pain, a common clinical problem.

8. Tail Incision

The tail is one organ that has been used widely to study pain-related behaviors. An advantage to studying the tail is the wide variety of nociceptive tests that can be utilized. For example, radiant heat, hot water, mechanical pinch, and electrical stimulation have been studied. The end point for tail tests is a sudden movement or flick of the tail. A decrease in latency, force, or threshold constitutes enhanced nociception; an increase in the latency, force, or threshold constitutes analgesia. Since the tail is readily accessible for pain behavior tests, it is not surprising that investigators developed a model of incisional pain using the tail of the rat.

A model of incisional pain has been developed by Weber and colleagues (26) using the tail of Sprague Dawley rats. Experiments are performed on adult rats that are anesthetized with 6% halothane and maintained with 2% halothane through a nose cone. The tails of the animals to undergo incision are prepared with chlorhexidine (Hibitane, Astra Zeneca), and the midpoint of the dorsal surface of the tail is marked. Incisions are made longitudinally on the tail over the midpoint, with half the incision above and half below the midpoint of the tail. Rats undergo two kinds of incisions: 10-mm longitudinal incision through skin and fascia or 20-mm incision through skin, fascia, and muscle. Briefly, in the first group of rats, a 10-mm longitudinal incision is made through skin and fascia on the dorsal surface of the center of the tail with a number 15 scalpel blade. The incision is sutured once with 4-0 nylon. The rats are allowed to recover in their cages for 90 min after surgery. In a second group of rats, a 20-mm longitudinal incision is made through skin, fascia, and muscle on the dorsal surface of the center of the tail, and suture with two 4–0 nylon sutures. All other procedures are identical to those performed on the rats with 10-mm skin and fascia incisions. The incisions are sprayed with a topical antiseptic (Necrospray, Centaur Labs, South Africa) and wrapped in protective film (Tegaderm, 3M Healthcare, USA) on the day of surgery, and the protective film is replaced every second day after surgery until the end of the study.

For testing the tail, rats are usually restrained in a plexiglass holder, and the tail protrudes through an opening in the restraining device. The tail withdrawal latency to noxious mechanical or heat stimuli is examined at the site of incision and proximal to the wound to test for hyperalgesia. In order to examine primary mechanical hyperalgesia, a bar algometer is placed directly onto the site of incision with a force of 4 N. The withdrawal latency is recorded, and the test is repeated three times on each animal with 1 min between tests. Secondary mechanical hyperalgesia is measured with the bar algometer placed 15 mm proximal to the incision with a force of 4 N. Withdrawal latency is recorded three times for each rat, at 1 min intervals. Testing is terminated if the rat has not reacted after 30 s.

To test the tail withdrawal latency to noxious heat, the rats are restrained and allowed to rest with their tails submerged in 29°C water for 30 min before the test. After this time, the whole tail is immersed in 49°C water, a noxious heat challenge. Withdrawal latency is recorded three times for each rat, at 1 min intervals. The test is terminated if the rat has not reacted after 30 s.

The average of the three withdrawal latencies for either heat or mechanical stimuli is recorded as the withdrawal latency for that day. The preincision value is taken to be the average of the withdrawal latencies of the 3 days of preincision recordings. Because the reaction times differ between rats, the percentage change, instead of response latency in seconds is used to compare preincision values to subsequent ones. The percentage change from preincision values is calculated for each rat for every day of testing using the formula: $[(\text{daily average value} - \text{pretrail value})/\text{pretrail value}] \times 100$.

The changes in pain behaviors observed from this tail incisional model indicate that primary mechanical hyperalgesia persisted for 6 days after surgery in rats with 10-mm skin and fascia incisions (27). In the case of the 20-mm skin, fascia, and muscle incision, primary mechanical hyperalgesia was evident for 7 days postoperatively. Similarly, secondary mechanical hyperalgesia was of 1 day longer duration following the 20-mm skin, fascia, and muscle incision than after the 10-mm skin and fascia incision. The results also show that primary mechanical hyperalgesia was of greater intensity for 3 days after the more extensive incision. Heat hyperalgesia was not evident in either incision.

The tail incisional model produces mechanical hyperalgesia immediately after incision; it sustains for about 6 days and represents a time course similar to that seen after minor surgical interventions in humans. Both primary and secondary mechanical hyperalgesia occur.

9. Conclusion

Our understanding of pathophysiologic mechanisms provides the rationale for future therapies in pain medicine. By developing models that examine the etiology and novel treatments of pain caused by surgery, new efficacious therapies will occur. The goal will be to nearly eliminate perioperative pain, decrease hospital costs, and reduce morbidity.

References

1. Apfelbaum, J. L., Chen, C., Mehta, S. S., and Gan, T. J. (2003) Postoperative pain experience: results from a national survey suggest postoperative pain continues to be undermanaged. Anesth Analg 97, 534–40.
2. Ballantyne, J. C., Carr, D. B., Deferranti, S., Suarez, T., Lau, J., Chalmers, T. C., Angelillo, I. F., and Mosteller, F. (1998) The comparative effects of postoperative analgesic therapies on pulmonary outcome – cumulative meta-analyses of randomized, controlled trials. Anesth Analg 86, 598–612.
3. Brennan, T. J., Vandermeulen, E., and Gebhart, G. F. (1996) Characterization of a rat model of incisional pain. Pain 64, 493–501.
4. Siegel, S., and Castellan, N. J. (1988) Nonparametric Statistics for the Behavioral Sciences, 2nd ed., McGraw-Hill, Inc., New York.
5. Chaplan, S. R., Bach, F. W., Pogrel, J. W., Chung, J. M., and Yaksh, T. L. (1994) Quantitative assessment of tactile allodynia in the rat paw. J Neurosci Methods 53, 55–63.
6. Whiteside, G. T., Harrison, J., Boulet, J., Mark, L., Pearson, M., Gottshall, S., and Walker, K. (2004) Pharmacological characterisation of a rat model of incisional pain. Br J Pharmacol 141, 85–91.
7. Stubhaug, A., Breivik, H., Eide, P. K., Kreunen, M., and Foss, A. (1997) Mapping of punctate hyperalgesia around a surgical incision demonstrates that ketamine is a powerful suppressor of central sensitization to pain following surgery. Acta Anaesth Scand 41, 1124–32.
8. Zahn, P. K., and Brennan, T. J. (1999) Primary and secondary hyperalgesia in a rat model for human postoperative pain. Anesthesiology 90, 863–72.
9. Martinez, V., Fletcher, D., Bouhassira, D., Sessler, D. I., and Chauvin, M. (2007) The evolution of primary hyperalgesia in orthopedic surgery: quantitative sensory testing and clinical evaluation before and after total knee arthroplasty. Anesth Analg 105, 815–21.
10. Banik, R. K., Subieta, A. R., Wu, C., and Brennan, T. J. (2005) Increased nerve growth factor after rat plantar incision contributes to guarding behavior and heat hyperalgesia. Pain 117, 68–76.
11. Brennan, T. J., Zahn, P. K., and Pogatzki-Zahn, E. M. (2005) Mechanisms of incisional pain, in Anesthesia Clinics of North America (Joshi, G. P., and Fleisher, L. A., Eds.) pp. 1–20, Saunders, Philadelphia.
12. Duarte, A. M., Pospisilova, E., Reilly, E., Hamaya, Y., Mujenda, F., and Strichartz, G. R. (2005) Reduction of post-incisional allodynia by subcutaneous bupivacaine: findings with a new model in the hairy skin of the rat. Anesthesiology 103, 113–25.
13. Pogatzki, E. M., Niemeier, J. S., and Brennan, T. J. (2002) Persistent secondary hyperalgesia after gastrocnemius incision in the rat. Eur J Pain 6, 295–305.
14. Pogatzki, E. M., Niemeier, J. S., Sorkin, L. S., and Brennan, T. J. (2003) Spinal glutamate

receptor antagonists differentiate primary and secondary mechanical hyperalgesia caused by incision. Pain 105, 97–107.
15. Beitz, A. J., Newman, A., Shepard, M., Ruggles, T., and Eikmeier, L. (2004) A new rodent model of hind limb penetrating wound injury characterized by continuous primary and secondary hyperalgesia. J Pain 5, 26–37.
16. Pitcher, G. M., Ritchie, J., and Henry, J. L. (1999) Nerve constriction in the rat: model of neuropathic, surgical and central pain. Pain 83, 37–46.
17. Lascelles, B. D., Waterman, A. E., Cripps, P. J., Livingston, A., and Henderson, G. (1995) Central sensitization as a result of surgical pain: investigation of the pre-emptive value of pethidine for ovariohysterectomy in the rat. Pain 62, 201–12.
18. Lascelles, B. D. X., Cripps, P. J., Jones, A., and Waterman, A. E. (1997) Post-operative central hypersensitivity and pain – the pre-emptive value of pethidine for ovariohysterectomy. Pain 73, 461–71.
19. Moiniche, S., Dahl, J. B., Erichsen, C.-J., Meinert Jensen, L., and Kehlet, H. (1997) Time course of subjective pain ratings, and wound and leg tenderness after hysterectomy. Acta Anaesthesiol Scand 41, 785–9.
20. Martin, T. J., Buechler, N. L., Kahn, W., Crews, J. C., and Eisenach, J. C. (2004) Effects of laparotomy on spontaneous exploratory activity and conditioned of operant responding in the rat – a model for postoperative pain. Anesthesiology 101, 191–203.
21. Martin, T. J., Kahn, W. R., and Eisenach, J. C. (2005) Abdominal surgery decreases food-reinforced operant responding in rats: relevance of incisional pain. Anesthesiology 103, 629–37.
22. Landreneau, R. J., Pigula, F., Luketich, J. D., Keenan, R. J., Bartley, S., Fetterman, L. S., Bowers, C. M., Weyant, R. J., and Ferson, P. F. (1996) Acute and chronic morbidity differences between muscle-sparing and standard lateral thoracotomies. J Thorac Cardiovasc Surg 112, 1346–50; discussion 1350–1.
23. Rogers, M. L., Henderson, L., Mahajan, R. P., and Duffy, J. P. (2002) Preliminary findings in the neurophysiological assessment of intercostal nerve injury during thoracotomy. Eur J Cardiothorac Surg 21, 298–301.
24. Nara, T., Saito, S., Obata, H., and Goto, F. (2001) A rat model of postthoracotomy pain: behavioural and spinal cord NK-1 receptor assessment. Can J Anaesth 48, 665–76.
25. Buvanendran, A., Kroin, J. S., Kerns, J. M., Nagalla, S. N., and Tuman, K. J. (2004) Characterization of a new animal model for evaluation of persistent postthoracotomy pain. Anesth Analg 99, 1453–60.
26. Weber, J., Loram, L., Mitchell, B., and Themistocleous, A. (2005) A model of incisional pain: the effects of dermal tail incision on pain behaviours of Sprague Dawley rats. J Neurosci Methods 145, 167–73.
27. Kamerman, P., Koller, A., and Loram, L. (2007) Postoperative administration of the analgesic tramadol, but not the selective cyclooxygenase-2 inhibitor parecoxib, abolishes postoperative hyperalgesia in a new model of postoperative pain in rats. Pharmacology 80, 244–8.
28. Wu, C., Erickson, M. A., Xu, J., Wild, K. D., and Brennan, T. J. (2009) Expression profile of nerve growth factor after muscle incision in the rat. Anesthesiology 110, 140–9.

Index

Chao Ma and Jun-Ming Zhang (eds.), *Animal Models of Pain*, Neuromethods, vol. 49, DOI 10.1007/978-1-60761-880-5, © Springer Science+Business Media, LLC 2011

U

V

W

Z

MIX
Papier aus verantwortungsvollen Quellen
Paper from responsible sources
FSC® C105338

If you have any concerns about our products,
you can contact us on
ProductSafety@springernature.com

In case Publisher is established outside the EU,
the EU authorized representative is:
Springer Nature Customer Service Center GmbH
Europaplatz 3, 69115 Heidelberg, Germany

Printed by Libri Plureos GmbH
in Hamburg, Germany